Springer Theses

Recognizing Outstanding Ph.D. Research

Aims and Scope

The series "Springer Theses" brings together a selection of the very best Ph.D. theses from around the world and across the physical sciences. Nominated and endorsed by two recognized specialists, each published volume has been selected for its scientific excellence and the high impact of its contents for the pertinent field of research. For greater accessibility to non-specialists, the published versions include an extended introduction, as well as a foreword by the student's supervisor explaining the special relevance of the work for the field. As a whole, the series will provide a valuable resource both for newcomers to the research fields described, and for other scientists seeking detailed background information on special questions. Finally, it provides an accredited documentation of the valuable contributions made by today's younger generation of scientists.

Theses are accepted into the series by invited nomination only and must fulfill all of the following criteria

- They must be written in good English.
- The topic should fall within the confines of Chemistry, Physics, Earth Sciences, Engineering and related interdisciplinary fields such as Materials, Nanoscience, Chemical Engineering, Complex Systems and Biophysics.
- The work reported in the thesis must represent a significant scientific advance.
- If the thesis includes previously published material, permission to reproduce this must be gained from the respective copyright holder.
- They must have been examined and passed during the 12 months prior to nomination.
- Each thesis should include a foreword by the supervisor outlining the significance of its content.
- The theses should have a clearly defined structure including an introduction accessible to scientists not expert in that particular field.

More information about this series at http://www.springer.com/series/8790

Yongkang Wu

Multi-physics Coupling Analysis of Clayey Core Wall of High Earth-Rockfill Dam

Doctoral Thesis accepted by
Tsinghua University, Beijing, China

Springer

Author
Dr. Yongkang Wu
State Key Laboratory of Hydroscience and Engineering, Department of Hydraulic Engineering
Tsinghua University
Beijing, China

Supervisor
Prof. Yuzhen Yu
State Key Laboratory of Hydroscience and Engineering, Department of Hydraulic Engineering
Tsinghua University
Beijing, China

ISSN 2190-5053 ISSN 2190-5061 (electronic)
Springer Theses
ISBN 978-981-13-1031-7 ISBN 978-981-13-1032-4 (eBook)
https://doi.org/10.1007/978-981-13-1032-4

Library of Congress Control Number: 2018945086

Printed on acid-free paper

This Springer imprint is published by the registered company Springer Nature Singapore Pte Ltd.
The registered company address is: 152 Beach Road, #21-01/04 Gateway East, Singapore 189721, Singapore

Supervisor's Foreword

It is my pleasure to introduce the Ph.D. thesis of Dr. Yongkang Wu for publication in the series of Springer Theses. Dr. Wu started his research as a Ph.D. student under my supervision in September 2011 and graduated in June 2016. During years of study at Tsinghua University, Dr. Wu has achieved many achievements and honors.

The work of this thesis focuses on the coupling analysis of seepage and consolidation problems in the clayey core wall of high earth-rockfill dams. In recent years, many high earth-rockfill dams with heights of 200–300 m have been built in China. For high earth-rock dams, the clayey core wall bears extremely complex changes in stress, deformation, and saturation during construction and operation, the seepage and consolidation behaviors of which become particularly important in engineering practice. This topic was studied and discussed thoroughly in this thesis. The mechanical properties of saturated–unsaturated compacted clayey soil were experimentally studied. New theories and methods for saturated–unsaturated consolidation analysis considering the interactions between the seepage, the stress, the deformation, the degree of saturation, etc., were developed. Moreover, an efficient and practical finite element program for large-scale complex earth-rockfill dams was developed and applied to the coupling analysis of seepage and consolidation in a high earth-rockfill dam. The theories and methods help to reveal the behaviors of compacted clayey soils under complex conditions of saturated–unsaturated, high stress, large deformation, and high pore pressure.

This thesis work shows new helpful ideas and perspectives for analyzing seepage and consolidation problems of high earth-rockfill dams as well as other projects in geotechnical engineering, and it will be of interest to scholars working on this topic.

Beijing, China
April 2018

Prof. Yuzhen Yu

Abstract

The seepage and consolidation behaviors of the clayey core wall in high earth-rockfill dams are significantly different from those in low dams, showing apparent multi-physics coupling effects. In this dissertation, to decipher the coupling mechanism, the mechanical properties of unsaturated compacted clayey soil, the unsaturated consolidation theory, and the multi-physics coupling analysis method are investigated, and an efficient and practical finite element (FE) program for large-scale complex high earth-rockfill dams is developed. The main work and contribution of this dissertation are as follows:

(1) Saturated and unsaturated triaxial consolidated-drained (CD) tests are conducted using the core material of gravelly clay from the Nuozhadu high earth-rockfill dam to reveal the mechanical properties of the core material in different saturation states. Due to the role of matric suction on the effective stress, the strength and modulus of the unsaturated soil are significantly higher than those of the saturated soil, especially under low confining pressures. The unsaturated test simulates the actual state of the field core material under rolling construction, and results indicate that the mechanical properties of the core material will change significantly with the degree of saturation during construction and impounding
(2) A multi-physics coupling analysis method for unsaturated seepage and consolidation problems is established based on a simplified consolidation theory to characterize the interactions between the physical-state field, the stress-deformation field, and the seepage field in the core wall. In this method, core issues, such as the calculation of effective stress, the modeling of soil permeability, and the constitutive models, are also emphatically clarified.
(3) Based on the established coupling analysis method and the derived finite element equations, a program for unsaturated seepage and consolidation analysis is developed. Moreover, efficient solvers for large-scale linear equations, the nonlinear iterative algorithm, and the under-relaxation method, are introduced into the program to improve the accuracy, efficiency, and computational stability. Numerical examples of the program show that the matrix suction has

significant effects on the stress and deformation behaviors of the dam. Compared with the traditional programs for saturated analysis, the program developed in this dissertation has significant advantages.

(4) The multi-physics coupling analysis program is applied to analyze the seepage and consolidation behaviors of the Nuozhadu high earth-rockfill dam. The simulated spatial and temporal distributions of the pore-water pressure in the core wall agree well with the field observation data, which implies the rationality of the coupling consolidation analysis method and the validity of the FE program. This program is suitable for large-scale three-dimensional consolidation analyses of high earth-rockfill dams.

Keywords High earth-rockfill dam · Multi-physics coupling · Consolidation Unsaturated · Clayey core wall

Contents

List of Figures

List of Tables

Chapter 1
Introduction

1.1 Background and Key Issues

1.1.1 Research Background

The earth-rock dam, as the oldest dam type, has many merits (Mai 2005) such as the availability of natural earth and rock materials, the ability to adapt to complex geological, topographic, and climatic conditions, and especially the ability to adapt to high-intensity earthquakes, making it one of the most commonly used dam types in hydropower projects. In recent years, earth-core rockfill dams account for a large proportion of high dams with a large reservoir in western China. Moreover, because of the growing social demand and the development of construction technology, earth-rockfill built in China are being built higher and higher.

Many of the dams under construction or planning were designed with heights of 200–300 m, or even higher (e.g., the Nuozhadu dam with a height of 261.5 m and the Rumei dam with a height of 315 m located on the Lancang River, the Shuangjiangkou dam with a height of 314 m located on the Dadu River, and the Lianghekou dam with a height of 292 m located on the Yalong River). The core wall is a key part of this kind of dams; it bears extremely complex changes in seepage and stress-deformation state in the processes of construction, impounding, and operation. In high earth-rock dams, the internal stress is high and large shear deformation exists, which result in significantly different physical states, stress-deformation behaviors, and seepage behaviors of the dam materials from those of low dams.

During construction and operation of several high earth-rockfill dams in China, the observed values of pore-water pressure, deformation, and their trends are quite different from the numerical results obtained by using the traditional consolidation analysis method. These phenomena are quite difficult to explain reasonably with previous theories and methods, which may be because the physical state, the stress-deformation, and the seepage have a complex coupling relationship during construction, impounding and operation; the engineering characteristics of

Y. Wu, *Multi-physics Coupling Analysis of Clayey Core Wall of High Earth-Rockfill Dam*, Springer Theses, https://doi.org/10.1007/978-981-13-1032-4_1

the unsaturated soil are very different from those of saturated soils. The traditional methods of consolidation analysis show obvious deficiencies, and more reasonable models and methods should be developed for analyzing and simulating the coupling relationship in high earth-rockfill dams. This research topic deals with the engineering characteristics of compacted clayey soils, the constitutive model, the multi-physics coupling consolidation analysis of saturated-unsaturated soils, and so on.

1.1.2 Key Issues

The core wall is a key part of the earth-rock dams; it bears extremely complex changes in seepage state and stress-deformation state in the process of construction, impounding, and operation. Due to the complexity of the construction conditions and the limitations of the test conditions and numerical methods, the existing numerical simulation analysis methods cannot adequately reflect the actual stress, deformation, and pore-water pressure states of the dam materials. Also, there are complex interactions between stress-deformation field, the physical-state field and the seepage field in the dam body. In particular, for the problems of saturated-unsaturated seepage and consolidation analysis in high earth-rock dams, the effect of multi-physics coupling should not be neglected.

During construction and operation of several high earth-rockfill dams, relatively high pore-water pressures have been observed in the core wall. The observed values of pore-water pressure and their trends are quite different from the estimated values obtained by numerical methods. For instance, the Xiaolangdi earth-rockfill dam in China (Chen et al. 2005), with a height of 160 m, was completed 12 years ago, but the pore-water pressure in the core wall still stays at a high level and changes periodically with the reservoir water level, and the water head in the core wall even exceeds the crest elevation. The Nuozhadu earth-rockfill dam was completed at the end of 2012, and the variations of pore-water pressure in the core wall have similar features. Some observed values of pore-water pressure in the core wall during construction and operation are one order of magnitude larger than the calculated values. These phenomena indicate the deficiencies of the previous analysis methods and theories. And the engineering properties of the compacted clayey soil and the analysis method of consolidation for high earth-rockfill dams need to be further studied.

Based on the observation data of the Nuozhadu dam and the Xiaolangdi dam, it is found that the existing models and numerical methods can hardly be verified by the field observation data of the high earth-rockfill dams. The numerical results of the stress, deformation, and pore-water pressure are significantly different from the observation data. For many completed high earth-rockfill dams, the actual dissipation rate of pore-water pressure in the core wall is significantly smaller than the values calculated by traditional methods. During the dam construction and a long time after completion, the pore-water pressure of the core wall is far greater than that in a steady-seepage period, and the water head in the core wall even exceeds the elevation of the dam crest. These phenomena are difficult to explain reasonably and simulate accurately with previous theories and methods. Thus, more reasonable models and methods should be developed for analyzing and simulating the coupling relationship in high earth-rockfill dams.

1.2 Scope of the Dissertation

For the engineering properties of the compacted clayey soil, saturated soil specimens are mostly used to study the hydraulic and mechanical properties of the soil. However, in practical high earth-rockfill dams, the degree of saturation of the core material after roller compaction is about 85%. The core material is unsaturated, and its mechanical and hydraulic properties are significantly different from those under saturated conditions. Moreover, the stress state and the degree of saturation of the core material will change significantly during construction and impounding, which has not been fully considered in the previous studies. Therefore, more reasonable test methods should be used to simulate the real changes in the mechanical properties of the core material during construction and operation.

For the consolidation analysis of earth-rockfill dams, the saturated consolidation theories have been developed to a relatively mature stage in both theoretical derivation and numerical simulation, and have been successfully adopted in the consolidation analyses of complex earth-rock dam projects. However, due to the restrictions of the basic assumptions in the saturated consolidation theories, they can hardly be used to analyze the unsaturated consolidation problems of soil. The theories show many deficiencies when analyzing the seepage and consolidation of high earth-rock dams. In view of this, the unsaturated consolidation theory should be adopted to analyze the related problems in earth-rockfill dams. At present, the unsaturated consolidation theories have got significant developments, but most of the theories sill stay in the stage of theoretical analysis. It is very complicated to apply them to solving practical problems, and thus there are very few practical application examples. Also, although the existing simplified consolidation theories for unsaturated soils can be applied in the analyses of simple problems, they still need to be further improved. For example, the rationality of the theoretical simplification needs to be verified, and the multi-physics coupling effects should be taken into account. During construction, impounding, and operation of high earth-rock dams, the compacted clayey soil of the core wall undergoes hugely complex changes in physical state, stress state, and seepage state. The spatial and temporal distribution of the physical state (e.g., density, degree of saturation, and microstructure) in the core wall is highly inhomogeneous, which will affect the spatial and temporal variations of soil permeability; the temporal and spatial nonuniformity of permeability will affect the stress and deformation distribution; and obviously, the change of mechanical state in soil will affect the physical state and the seepage characteristics. Hence, there are complex coupling relationships between the physical-state field, the seepage field, the deformation field, the stress field, and the surrounding environment. And the current theories and methods still hardly reflect these characteristics. In order to solve the above problems, based on the existing consolidation theories for unsaturated soils, a more reasonable saturated-unsaturated seepage and consolidation analysis method which can be used in solving large-scale complex engineering problems is proposed in this dissertation.

For the numerical solutions of the unsaturated seepage-consolidation analysis, poor numerical stability and low computational efficiency are the main problems due to the complex dam boundaries and the considerable differences in mechanical and permeability properties between different dam materials. It is much more difficult to solve the consolidation problems of unsaturated soil than to solve the saturated consolidation problems. On the one hand, the theory is relatively complicated with more parameters and variables; on the other hand, the material nonlinearity is more significant. In the development of the finite element (FE) program in this dissertation, the above problems of numerical solutions should be settled properly. To improve the computational efficiency, the efficient solutions for linear systems of equations which have been successfully used in other large-scale engineering problems can be introduced into the FE program. Moreover, to improve the numerical stability, more effective iterative methods for problems of material nonlinearity can also be adopted.

The above issues will be studied in depth in this study. The main contents of this dissertation include the following chapters.

In Chap. 2, the main research progress on unsaturated compacted clayey soils is summarized from the aspects of the engineering properties, the constitutive models, the strength theories, and the coupling analysis methods of seepage and consolidation. The related theoretical basis and the deficiencies of the current research results are also reviewed.

In Chap. 3, in order to reveal the strength and deformation properties of the core material under saturated and unsaturated (at optimal water content) conditions, triaxial consolidated-drained tests are carried out with the gravelly clay of the core material in the Nuozhadu high earth-rockfill dam. According to the test results, the strength and modulus differences between the unsaturated and the saturated soils under the same test conditions are analyzed.

In Chap. 4, a multi-physics coupling analysis method is introduced in the simplified consolidation theory for unsaturated soils to characterize the interactions of the physical state, the stress-deformation, and the seepage in the soil and a more reasonable multi-physics coupling analysis mode for seepage-consolidation is established. Moreover, the core issues in the multi-physics coupling analysis method (e.g., the constitutive model, the description of soil permeability, and the calculation of effective stress) are clarified.

In Chap. 5, based on the simplified consolidation theory and the coupling analysis method, more general FE equations for unsaturated seepage and consolidation analysis are derived, and a FE program for numerical simulations of multi-physics coupling unsaturated consolidation problems is developed. In order to improve the computational accuracy and efficiency and the numerical stability of the program, multiple types of elements, the iterative method for nonlinear problems, the under-relaxation method, and the iterative algorithms for large-scale linear systems of equations are introduced into the FE program. Moreover, two typical examples of earth-rock dams are analyzed to verify the analysis method as well as the FE program.

In Chap. 6, the field observation data of pore-water pressure in the core wall of the Nuozhadu high earth-rockfill dam is analyzed. Moreover, the FE program developed in this dissertation is applied to the analysis of seepage and consolidation of the dam. Comparing the numerical results with the observation data, the reasonability of the results is proved, and the multi-physics coupling mechanism in the core wall of high earth-rockfill dams is explained. Then the applicability of the theory and program established in this dissertation to the large-scale three-dimensional consolidation analyses of complex high earth-rockfill dams is discussed.

In Chap. 7, the research results, main conclusions, innovations, and future works of this dissertation are summarized.

References

Chen LH, Chen ZY, Zhang JP, Zhao C (2005) Study on high pore pressure in clay core wall of earth dam in Xiaolangdi Project. J Hydraul Eng 02:219–224

Mai J (2005) Hydraulic structures. Tsinghua University Press, Beijing

Chapter 2
Review of the Relevant Research Progress

The multi-physics coupling analysis in the core wall of high earth-rockfill dam involves a wide range of contents including the engineering properties of unsaturated soils, the unsaturated seepage-consolidation theory, the coupling analysis method, the theory of finite element method, the numerical solution algorithm, and so on. In this chapter, the main research progress on unsaturated clayey soils is summarized from the aspects of the stress-state variables, the strength theories, the constitutive models, and the coupling analysis methods of seepage and consolidation. The related theoretical basis and the deficiencies of the current research results are also reviewed.

2.1 Mechanical Properties of Compacted Unsaturated Clayey Soils

The studies on the mechanical Properties of saturated soils have been relatively in-depth and will not be discussed here. The following briefly introduces the research progress of the mechanical Properties of compacted unsaturated clayey soils from the aspects of the stress-state variables, the effective stress principle, the strength characteristics, and the deformation characteristics.

2.1.1 Stress-State Variables

2.1.1.1 Single Stress-State Variable Theory

A reasonable description of the stress state of a soil is the basis for the study of its mechanical properties. The stress state in soil can be expressed in different ways. For saturated soils, the concept of effective stress has been widely accepted. The effective

Y. Wu, *Multi-physics Coupling Analysis of Clayey Core Wall of High Earth-Rockfill Dam*, Springer Theses, https://doi.org/10.1007/978-981-13-1032-4_2

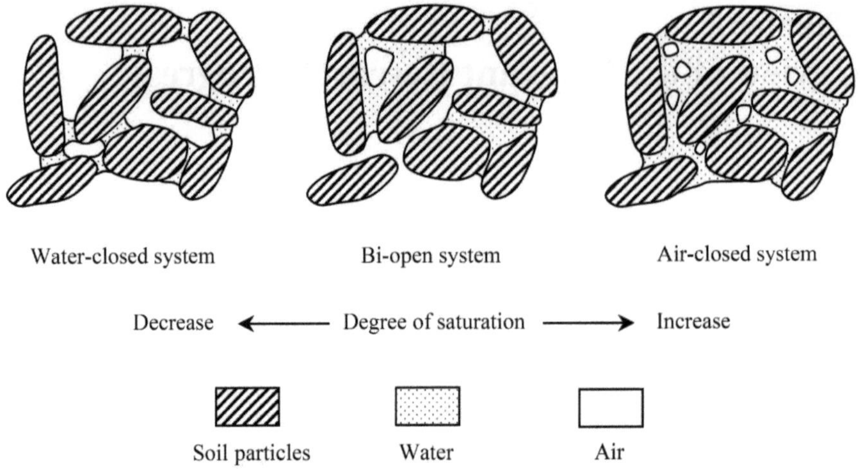

Fig. 2.1 The water-air morphologies in unsaturated soils

stress principle for saturated soils, which is the basis of traditional soil mechanics, was firstly developed by Terzaghi. The effective stress, σ', in a saturated soil can be expressed in terms of the total stress, σ, and the pore-water pressure, u_w, as follows:

$$\sigma' = \sigma - u_w \tag{2.1}$$

The deformation and strength behaviors of saturated soils can be described by the effective stress. For unsaturated soils, the description of stress state is much more complicated than that of saturated soils. And a lot of explorations have been made in determining the variables of stress state in unsaturated soils.

Yu and Chen (1965) classified the unsaturated soils into three different states based on the water-air morphology in the soils, as shown in Fig. 2.1. When a soil's degree of saturation is low, the pore space in the soil is mainly occupied by air, and the air in the pores will be interconnected. In this case, the pore water is surrounded by the pore air and in a "closed" state. Hence, the unsaturated soil is called a water-closed system. When the soil's degree of saturation is relatively high, the pore water is interconnected, and the pore air will be enclosed in the water. In this case, the soil is called an air-closed system. When the soil's degree of saturation is between those of the first two states, both the air phase and the water phase in the soil are interconnected. Then, the soil is called a bi-open system.

In particular, for the air-closed system and water-closed system, the unsaturated soil can be approximated as a saturated soil and a dry soil, respectively. And correspondingly, the effective stress can be expressed by Eq. (2.1) and the following equation:

$$\sigma' = \sigma - u_a \tag{2.2}$$

where u_a is the pore-air pressure.

For general cases, scholars have proposed various types of effective stress equations for unsaturated soils, of which the most representative one is the Bishop's equation proposed in 1959 (Bishop 1959):

$$\sigma' = \sigma^* + \chi s \tag{2.3}$$

The above equation can also be expressed as follows:

$$\sigma' = \sigma - [\chi u_w + (1 - \chi)u_a] \tag{2.4}$$

where, $\sigma^* = \sigma - u_a$ is the net stress; $s = u_a - u_w$ is the matric suction; χ is the effective stress parameter, which is used to describe the contribution of matric suction to the effective stress, also known as the reduction coefficient or the suction parameter; $\chi(u_a - u_w)$, also indicated by the symbol $\bar{s}$, is the reduced matric suction (also known as the equivalent suction). The effective stress parameter of a soil is not a constant and will change with the change of the matric suction or the degree of saturation. The relationship between the effective stress parameter, χ, and the degree of saturation, S_r, of soil has been experimentally investigated by many scholars (Blight 1961; Donald 1961; Escario et al. 1989).This equation involves two special situations of saturated soil and dry soil. When the soil is saturated, $\chi = 1$, and the effective stress equation will evolve into the effective stress equation of saturated soils in Eq. (2.1). When the soil is completely dry, $\chi = 0$.

Aitchison (1961) and Jennings (1961) also gave similar expressions for the effective stress of unsaturated soils. However, due to the limitations of Bishop's effective stress equation (e.g., the lack of reasonable explanations of mechanism, and the inability to describe the phenomenon of collapsibility of unsaturated soils), the application of the equation has been significantly limited. Bao (2004) also pointed out that this principle failed to fully reflect the influence of the air phase (e.g., the suction, the contractile skin, etc.) on the mechanical properties of unsaturated soils.

In the earlier studies, it was generally assumed that the effective stress parameter is only related to the degree of saturation. Among these studies, Mctigue et al. (1983) and Nikolaevsky (1985) directly substituted the degree of saturation, S_r, for the effective stress parameter, χ, and defined the effective stress as follows:

$$\sigma'_{ij} = \sigma_{ij} - [S_r u_w + (1 - S_r)u_a]\delta_{ij} \tag{2.5}$$

The above equation which is also known as the simplified Bishop formula is essentially a special case of the Bishop's effective stress formula.

Vardoulakis and Beskos (1986) proposed the following form of the effective stress equation:

$$\sigma_{ij} = \sigma^s_{ij} + S_r u_w \delta_{ij} + (1 - S_r)u_a \delta_{ij} \tag{2.6}$$

$$\sigma'_{ij} = \sigma^s_{ij} - \gamma u_w \delta_{ij} \tag{2.7}$$

where γ is the soil parameter; σ^s_{ij} is the stress between soil particles. However, this formula lacks theoretical basis and experimental support, and the meaning of the parameter is also unclear.

However, many scholars pointed out that the relationship between the effective stress parameters, χ, in Bishop's equation and the degree of saturation, S_r, is not single-valued. For example, Chen et al. (1994) studied the effective stress from the viewpoint of volume change in unsaturated soils based on the elastic theory, and proposed that for an unsaturated soil with the saturation of S_r and the porosity of n, the effective stress can be expressed by the soil bulk modulus, K:

$$\sigma' = \sigma - \left[\frac{K^n}{K^{S_r n}} u_w + \left(1 - \frac{K^n}{K^{S_r n}}\right) u_a\right] \tag{2.8}$$

where K^n and $K^{S_r n}$ are the bulk modulus of soil with the porosity of n and $S_r \cdot n$, respectively. Comparing Eq. (2.8) with the Bishop's effective stress equation yields:

$$\chi = \frac{K^n}{K^{S_r n}} \tag{2.9}$$

Since $K^n \leq K^{S_r n}$, it can be concluded that $0 \leq \chi \leq 1$. Accordingly, Chen deduced that the effective stress parameter, χ, is not only related to the degree of saturation but also to the void ratio of the soil. Khalili and Khabbaz (1995) and Chen and Chen (1999) also obtained the same conclusions.

Many of the above expressions of effective stress in unsaturated soils are similar to the Bishop's equation. Although these equations improved the Bishop's equation from different perspectives, they still did not deviate from the basic idea of the Bishop's theory. Hence, they do not have significant advantages over the Bishop's equation. Moreover, the Bishop's effective stress theory has many merits. For example, the theory is simple, its expression of effective stress is similar to that of saturated soils, and thus it is easy to use. Therefore, the Bishop's theory has been accepted by many scholars and got further developments (Kohgo et al. 1993; Modaressi and Abou-Bekr 1994; Shen 1995; Khalili and Khabbaz 1998).

2.1.1.2 Theory of Double Stress-State Variables

The theory of double stress-state variables was first proposed by Coleman (1962). He suggested that the "reduced" stress variables, $(\sigma_1 - u_a)$, $(\sigma_3 - u_a)$, and $(u_a - u_w)$, can be used to describe the stress states in triaxial tests. Bishop and Blight (1963) proposed a similar expression, in which the stress-state variables are expressed as follows:

$$\sigma^* = (\sigma - u_a), \quad s = (u_a - u_w) \tag{2.10}$$

Fredlund argued that unsaturated soils consist of four phases: the soil particles, the air, the water, and the air-water interface (i.e., the contractile skin). Based on this opinion, he derived the equilibrium equation of soil structure, and pointed out that without considering the compressibility of soil particles and water, the stress state that controls the balance of the soil structure and contractile skin can be represented by two independent stress-state variables (Fredlund and Rahardjo 1993). In 1977, Fredlund et al. verified the validity of the two stress-state variables and proposed that the net stress, $(\sigma - u_a)$, and the matric suction, $(u_a - u_w)$, can be used as the stress-state variables for mechanical problems, such as deformation analysis and strength analysis of unsaturated soils (Fredlund and Morgenstern 1977; Fredlund 1979). This theory has been accepted by many scholars.

In the 1990s, Chen (1994) studied the stress-state variables of unsaturated soils and proved the rationality of the double-stress variables theory proposed by Fredlund. Chen and Qin (2012) pointed out that the Fredlund's double stress-state variables belong to a simplified form because the factors of the degree of saturation, S_r, and the porosity, n, are not considered.

Bolzon et al. (1996) replaced the effective stress parameter, χ, in the Bishop's equation with the degree of saturation, S_r, and used the modified effective stress equation (also known as the simplified Bishop's equation) and the matric suction, s, as two stress-state variables for establishing the elasto-plastic constitutive relationship of unsaturated soils, as follows:

$$\sigma'_{ij} = \sigma_{ij} - u_a\delta_{ij} + S_r(u_a - u_w)\delta_{ij} \tag{2.11}$$

$$s = u_a - u_w \tag{2.12}$$

Houlsby (1997) was inspired by Eqs. (2.11) and (2.12) and proposed that the simplified Bishop's equation and ns (where n is the porosity of soil) can be regarded as the two stress-state variables of unsaturated soils.

Zhao et al. (2010) also obtained similar conclusions and proposed that $n(1 - S_r)u_a$ should be taken as the third stress-state variable of unsaturated soil when the pore-air is in a closed state. Hence, the stress-state variables are as follows:

$$\sigma'_{ij} = \sigma_{ij} - u_a\delta_{ij} + S_r(u_a - u_w)\delta_{ij} \tag{2.13}$$

$$s' = n(u_a - u_w) \tag{2.14}$$

$$u'_a = n(1 - S_r)u_a \tag{2.15}$$

In the above three combinations of stress-state variables proposed by Bolzon et al. (1996), Houlsby (1997), and Zhao et al. (2010), the factors of porosity or degree of saturation were taken into account. However, the matrix suction was repeatedly considered in these stress-state variable combinations (Chen 2014). The stress-state variables are not independent of each other, and they lack sufficient theoretical and experimental basis.

Considering that the matrix suction is affected by the surface tension and the physicochemical actions, Lu (2008) used a soil parameter X to reflect the influence of these factors on the mechanical properties of unsaturated soils. That is, $X(u_a - u_w)$ was chosen as one of the stress state variables.

Although many different combinations of stress-state variables have been proposed, the Fredlund's theory of double stress-state variables is still the most important one because of its wide range of applications and abundant results. Compared with the theory of effective stress, the theory of double stress-state variables has less limitations and more definite physical meanings. But it is very complicated to use in practical application. The theories in saturated soil mechanics cannot be employed any more, and it is necessary to reestablish the theories of strength, deformation, and consolidation for unsaturated soils.

2.1.1.3 Comparison of the Two Kinds of Theories

At present, there are mainly two ways to describe the stress state of unsaturated soils: the single stress-state variable and the double stress-state variables. For the theories of single stress-state variable, the Bishop's theory of effective stress is the most representative; for the theories of double stress-state variables, the Fredlund's theory is the most widely used. Both of the two kinds of theories can be adopted to describe the stress state in unsaturated soils.

Although there are significant differences between the theories of single stress-state variable and the theories of double stress-state variables, the two kinds of theories are also closely interrelated. For example, the effective stress with a single variable can be regarded as a comprehensive reflection of different stress state variables.

Chen (1994) compared the expressions of constitutive relationship and strength of the two kinds of theories and concluded that the theories are equivalent except that the methods for determining the parameters are different. Shen (2000) also pointed out that for non-collapsible soils under certain conditions, the research ideas of the two kinds of theories are interlinked. In addition, the basic assumptions of both kinds of theories are the same. Neither of them considered the compressibility of soil particles and water and the variation of internal friction angle with the degree of saturation change. However, there are also significant differences between them. Bishop's effective stress theory can't describe the collapsibility of soils, but its concept is simple and can follow the related research results in saturated soil mechanics; the Fredlund's theory of double stress-state variables has fewer limitations, but it is too complicated to apply in practice.

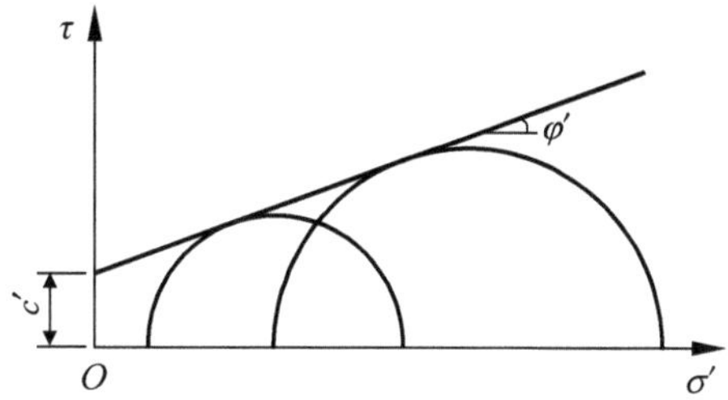

Fig. 2.2 Mohr-Coulomb failure envelope for saturated soils

2.1.2 Strength Theory

The shear strength of a saturated soil is described using the Mohr-Coulomb failure criterion:

$$\tau_f = c' + \sigma' \tan \varphi' \tag{2.16}$$

where τ_f is the shear strength; σ' is the effective stress; c' is the effective cohesion; φ' is the effective angle of internal friction. The strength equation in Eq. (2.16) defines a failure envelope, as illustrated in Fig. 2.2.

As with the theories of stress-state variables used to describe the stress state of unsaturated soils, the strength equations of unsaturated soils are also divided into two types: the strength equations of single stress-state variable and the strength equations of double stress-state variables. Of the two types of strength equations, the Bishop's strength equation and Fredlund's strength equation are the most representative equations, respectively (Hu et al. 2008).

According to the Bishop's equation of effective stress, Bishop expressed the strength of unsaturated soils as follows (Bishop 1959):

$$\tau_f = c' + [(\sigma - u_a) + \chi(u_a - u_w)] \tan \varphi' \tag{2.17}$$

Equation (2.17) has the same form as the strength equation for saturated soils in Eq. (2.16). The strength parameters of the equations include both the effective cohesion and the effective angle of internal friction. The only difference between them is the expressions of the effective stress. Therefore, the failure envelope of the Bishop's strength equation can also be described by Fig. 2.2. If the Bishop's strength equation is rearranged into the following form,

$$\tau_f = c' + (\sigma - u_a) \tan \varphi' + \chi(u_a - u_w) \tan \varphi' \tag{2.18}$$

the envelope of unsaturated soils can be further plotted in three-dimensional coordinates formed by the shear stress, τ, the net normal stress, $(\sigma - u_a)$, and the reduced suction, $\chi(u_a - u_w)$, as shown in Fig. 2.3. In the figure, $c'' = \chi(u_a - u_w) \tan \varphi'$, which represents the shear strength generated by the matric suction.

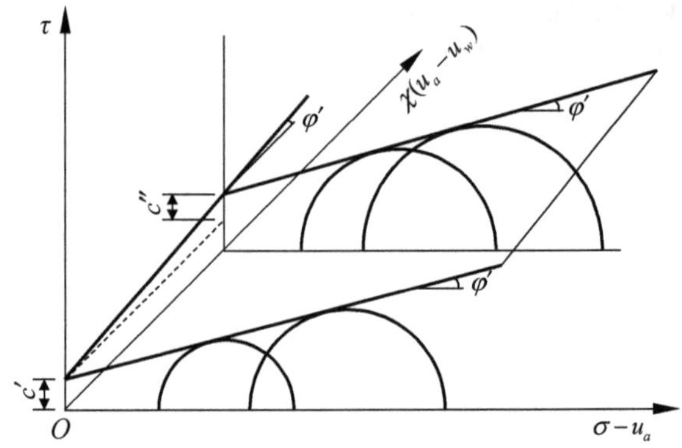

Fig. 2.3 Failure envelope for unsaturated soils (Bishop's strength equation)

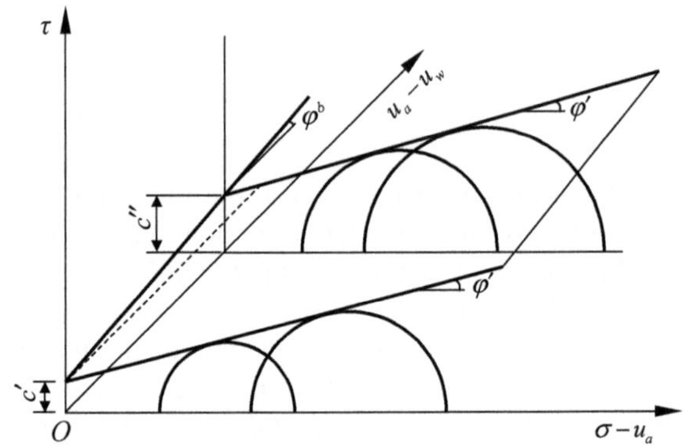

Fig. 2.4 Failure envelope for unsaturated soils (Fredlund's strength equation)

On the basis of the double stress-state variables theory, Fredlund et al. (1978) defined another form of strength equation for unsaturated soils using the net normal stress and the matric suction, the failure envelope of which are shown in Fig. 2.4:

$$\tau_{\mathrm{f}} = c' + (\sigma - u_a)\tan\varphi' + (u_a - u_w)\tan\varphi^{\mathrm{b}} \tag{2.19}$$

where φ' is the internal fraction angle associated with the net normal stress; φ^{b} is the slope of the failure envelope along with the axis of matric suction, indicating the change rate in shear strength relative to the matric suction.

A comparison of Eqs. (2.18) and (2.19) reveals that $\tan\varphi^{\mathrm{b}} = \chi\tan\varphi'$, and there is no substantial difference between the two equations. Because of the difficulties in determining the parameters of the above strength equations for unsaturated soils, many scholars have studied the shear strength equations from the practical point of view.

Based on the experimental results, Khalili and Khabbaz (1998) fitted the relationship between the effective stress parameter in the Bishop's equation and the matric suction:

$$\chi = \left[\frac{u_a - u_w}{(u_a - u_w)_b} \right]^{-0.55} \tag{2.20}$$

where $(u_a - u_w)_b$ is the air-entry pressure of soil. Then substituting the effective stress parameter of the Bishop's strength equation by Eq. (2.20) yields:

$$\tau_f = c' + \left[\sigma - u_a + (u_a - u_w)^{0.45} (u_a - u_w)_b^{-0.55} \right] \tan \varphi' \tag{2.21}$$

Vanapalli et al. (1996) proposed a strength equation related to the water content of soil:

$$\tau_f = c' + (\sigma - u_a) \tan \varphi' + (u_a - u_w) \left(\frac{\theta - \theta_r}{\theta_s - \theta_r} \right) \tan \varphi' \tag{2.22}$$

where θ is the volumetric water content; θ_s is the saturated volumetric water content; θ_r is the residual volumetric water content.

Öberg and Sällfors (1997) replaced the effective stress parameter of the Bishop's equation by the degree of saturation and proposed the following strength equation:

$$\tau_f = c' + (\sigma - u_a) \tan \varphi' + (u_a - u_w) S_r \tan \varphi' \tag{2.23}$$

Huang and Wu (2007) also proposed the same strength equation.

Hu et al. (2008) verified the two strength equations in Eqs. (2.22) and (2.23) using specific soil materials and pointed out that the strength equation proposed by Vanapalli is more reasonable than the other and the difference between the two equations is mainly reflected in the contributions of matric suction to shear strength.

Ling and Yin (2007) studied the shear-strength variation of unsaturated soils with the change of water content, w, using an improved triaxial apparatus, and concluded that the shear strength of unsaturated soils decreases with the increase of water content. Then by introducing the strength parameter functions related to the water content, the following strength equation for unsaturated soils was obtained:

$$\tau_f = c(w) + \sigma \tan \varphi(w) \tag{2.24}$$

where $c(w)$ and $\varphi(w)$ can be determined by linear fitting based on a set of tests with different water contents.

Shen et al. (2007) measured the effective strength parameters of clayey soils with different water contents and dry densities by traditional triaxial tests, and fitted the relationships between the effective strength parameters and the factors of water content and dry density.

Bian and Wang (2011) also studied the relationship between the shear strength of unsaturated soils and the water content and proposed the following equations to represent the total cohesion, c, and the internal friction angle, φ, respectively:

$$c(w) = a_1 w^2 + a_2 w + a_3 \tag{2.25}$$

$$\varphi(w) = b_1 w + b_2 \tag{2.26}$$

where a_1, a_2, a_3, b_1, and b_2 are fitting parameters.

2.2 Constitutive Models for Compacted Unsaturated Clayey Soils

Various constitutive models have been proposed to describe the stress-strain relationship of unsaturated soils. These constitutive models can be roughly divided into the several categories: the elastic models, the elasto-plastic models, the constitutive models for special soils, etc. The following briefly summarizes the commonly used elastic and elasto-plastic models in the unsaturated soil mechanics.

2.2.1 *Elastic Model*

Based on the generalized Hook's law, Fredlund and Morgenstern (1976) established the elastic relationship of unsaturated soils by using the net normal stress and the matric suction as the stress-state variables. The normal stress-stain relationship is expressed as follows:

$$\begin{cases} \varepsilon_x = \dfrac{(\sigma_x - u_a)}{E} - \dfrac{\mu}{E}(\sigma_y + \sigma_z - 2u_a) + \dfrac{(u_a - u_w)}{H} \\ \varepsilon_y = \dfrac{(\sigma_y - u_a)}{E} - \dfrac{\mu}{E}(\sigma_x + \sigma_z - 2u_a) + \dfrac{(u_a - u_w)}{H} \\ \varepsilon_z = \dfrac{(\sigma_z - u_a)}{E} - \dfrac{\mu}{E}(\sigma_x + \sigma_y - 2u_a) + \dfrac{(u_a - u_w)}{H} \end{cases} \tag{2.27}$$

where E is the Young's modulus of the unsaturated soil; μ is the Poisson's ratio of the unsaturated soil; H is the elastic modulus of the soil associated with the matric suction.

Yang (1990) proposed a nonlinear elastic model for unsaturated soils. Chen et al. (1998, 1999) also derived a nonlinear model for unsaturated soils based on the theory of double stress-state variables, and expressed the elastic parameters by functions of matric suction. This model can be regarded as an extension of the Duncan and Chang's EB model in unsaturated soils. That is, the parameters of K and K_b in the Duncan and Chang's EB model are regarded as linear functions of the matric suction, and the stress in the model is replaced by the net stress:

$$K = K^0 + c_1 \frac{s}{p_a} \tag{2.28}$$

$$K_b = K_b^0 + c_2 \frac{s}{p_a} \tag{2.29}$$

where c_1 and c_2 are parameters; K^0 and K_b^0 take the values of K and K_b of a soil when saturated, respectively. In this model, the Fredlund's strength equation was adopted as the strength equation, and the tangent modulus, E_t, can be expressed as:

$$E_t = K p_a \left[1 - \frac{R_f (1 - \sin \varphi')(\sigma_1 - \sigma_3)}{2(c' + s \tan \varphi^{\mathrm{b}}) \cos \varphi' + 2(\sigma_3 - \sigma_a) \sin \varphi'} \right] \tag{2.30}$$

Shen (2000) proposed that the constitutive model of saturated soils can also be used to describe the stress-strain relationship of unsaturated soils by replacing the stress in the saturated constitutive models with the effective stress defined by the effective stress equation of unsaturated soils. In this way, the Duncan and Chang's Eν model for saturated soils can be extended to unsaturated soils, and the initial modulus of E_i and ν_i can be expressed as follows:

$$E_i = K p_a \left(\frac{\sigma_3^* + \chi s}{p_a} \right)^n \tag{2.31}$$

$$\nu_i = G - F \lg \left(\frac{\sigma_3^* + \chi s}{p_a} \right) \tag{2.32}$$

where the Bishop's strength equation shown in Eq. (2.17) was adopted as the shear strength equation.

Yin et al. (2006) argued that the degree of saturation or the water content could also be used as a state variable instead of the matric suction in the constitutive models of unsaturated soils, and concluded the advantages and disadvantages of these two treatment methods.

Ling et al. (2008) proved through experiments that the stress-strain relationship of unsaturated soils could be described by the Duncan and Chang's model, and pointed out that the model parameters are functions of the water content.

2.2.2 *Elasto-plastic Models*

One of the most famous elasto-plastic models for unsaturated soils is the Barcelona model proposed by Alonso et al. (1990). This model introduced a component of matric suction into the stress space and extended the elliptical yield surface on the meridian plane of the modified Cam-clay Model to the yield surface in the three-dimensional stress space of *p-q-s*. Then, many scholars made further improvements to this elasto-plastic constitutive model (Chiu and Ng 2003).

Huang et al. (2000) adopted a new equation and combined the two yield lines, i.e., LC (loading-collapse) and SI (suction-increase), into a single yield surface.

However, the above constitutive models for unsaturated soils considered only the prediction of deformation and strength and neglected the influence of the water-phase change in unsaturated soils. In view of this, Wheeler et al. (2003), Miao (2007), and Sun et al. (2007) established elasto-plastic constitutive models with the coupling of hydraulic and mechanical properties of unsaturated soils. And based on the theory of multiphase porous media, Zhao and Liu (2009) deduced the elasto-plastic constitutive equations of unsaturated soils by using three generalized stresses as the independent stress-state variables.

2.3 Hydraulic Properties of Compacted Unsaturated Clayey Soils

The water retention characteristic of soil is commonly described by a soil-water characteristic curve (SWCC). The soil-water characteristic curve is a fundamental constitutive relationship in the unsaturated soil mechanics, which describes the relationship between the soil water content (or degree of saturation) and the soil matric suction, as shown in Fig. 2.5. The soil-water characteristic curve is closely related to the mechanical, seepage, and water-retention characteristics and is a key indicator of the unsaturated soil properties.

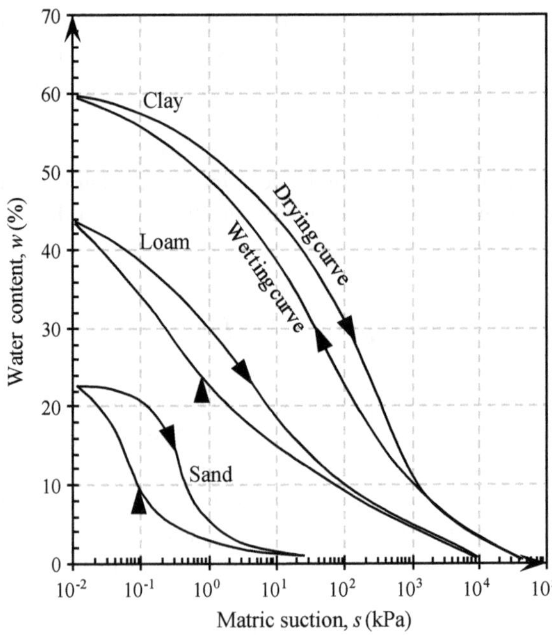

Fig. 2.5 Soil-water characteristic curve

The soil-water characteristic curve is affected by multiple factors, such as soil type, mineral composition, soil structure, compaction energy, water content, void ratio, dry density, stress history, and stress state (Vanapalli et al. 1999; Alonso and Cardoso 2010; Wang et al. 2009; Miller et al. 2002; Pereira and Fredlund 2000). Scholars have proposed various types of soil-water characteristic curve models (Gardner 1958; Brooks and Corey 1964; van Genuchten 1980; Williams et al. 1983; McKee and Bumb 1987; Fredlund et al. 1994). For geotechnical engineering applications, the Brooks-Corey (1964) model, the van Genuchten (1980) model, and the Fredlund-Xing model (Fredlund et al. 1994) are most commonly adopted.

The Brooks-Corey model, as one of the earliest models to describe soil-water characteristic curves, was proposed by Brooks and Corey in 1964 on the basis of a large amount of experimental data. The model is shown in Eq. (2.33). In the model, S_e is the effective degree of saturation; λ is a parameter reflecting the pore-size distribution of soil, and s_e is the air-entry pressure of soil. It can be seen that the soil-water characteristic curve described by this model can be divided into two parts at the point of air-entry pressure. The curve is not smooth, which will cause the numerical instability in seepage analysis.

$$\begin{cases} S_e = (s_e/s)^{\lambda} & s \geq s_e \\ S_e = 1 & s < s_e \end{cases} \tag{2.33}$$

The van Genuchten model, as shown in Eq. (2.34), was proposed in 1980. It contains a total of three parameters: α, n, and m. This model can describe the smooth transition and the shape of soil-water characteristic curve very well and has been widely used in unsaturated seepage analyses in geotechnical engineering.

$$S_e = \left[1 + (\alpha \cdot s)^n\right]^{-m} \tag{2.34}$$

The Fredlund-Xing model was proposed in 1994. The model can be expressed as follows:

$$\theta = C(s)\theta_s \left[\frac{1}{\ln[\mathrm{e} + (s/a)^n]}\right]^m \tag{2.35}$$

where e is the base of natural logarithms; α, n, and m are the fitting parameters of the model; and $C(s)$ is a correction factor, which can be expressed as:

$$C(s) = \left[1 - \frac{\ln(1 + s/s_r)}{\ln\left(1 + 10^6/s_r\right)}\right] \tag{2.36}$$

where s_r is the matric suction estimated at the residual condition.

Another important hydraulic characteristic parameter of soil is the hydraulic conductivity. For the seepage and consolidation analysis of both saturated soils and unsaturated soils, the value of hydraulic conductivity will significantly affect the whole process of seepage and consolidation. Thus, the determination of hydraulic conductivity becomes one of the core problems to be solved in seepage and consolidation analysis.

The permeability of saturated clayey soils is significantly influenced by the physical and mechanical states. Currently, scholars have conducted plenty of research on the seepage characteristics of saturated soils, studied the influence of various factors on the permeability, and proposed a series of hydraulic conductivity models. In terms of physical state, the effects of particle size, gradation, density, void ratio, plasticity index, etc., of soil on its permeability have been studied (Marshall 1958; Mesri and Olson 1971; Samarasinghe et al. 1982; Liu 1992; Yin 2009). And in terms of mechanical state, the effects of confining pressure, effective stress, degree of consolidation, shear stress level, etc., on the permeability of soil have also been studied (Carpenter and Stephenson 1986; Chu et al. 2002; Lei et al. 2016).

For unsaturated soils, the seepage characteristic is much more complicated than that of saturated soils. The hydraulic conductivity of unsaturated soils can be deduced indirectly based on the soil-water characteristic curve and the saturated permeability properties (Vanapalli et al. 1999; van Genuchten 1980; Leong and Rahardjo 1997). Although experimental measurement method is a more reliable way to obtain the hydraulic conductivity of unsaturated soils, it is very difficult to implement.

The hydraulic conductivity of unsaturated soils is affected by many factors. Huang et al. (1998) established a hydraulic conductivity equation for unsaturated soils under deformation conditions and verified the equation by triaxial seepage tests. Li et al. (2004) studied the influence of water content and confining pressure on soil permeability by using a triaxial seepage test device for unsaturated soils. Gao et al. (2005) studied the effects of dry density, degree of saturation, and wetting path on the permeability of water-phase and air-phase in loess using an apparatus for measuring water-air movement. Zhang et al. (2010) established a hydraulic conductivity model considering the influence of pore structure and simulated the effects of deformation on the permeability of unsaturated soils.

2.4 Coupling Analysis Theories of Seepage and Consolidation for Earth-Rock Dams

The coupling analysis of stress field and deformation field has already been a hot topic in the research area of the fractured rock mass. And there have been many mature results in experiments and numerical simulations (Ye and Liu 2005; Chang et al. 2004; Liu et al. 2002a, b; Wang et al. 2000; Chai 2000; Su et al. 1994; Barton et al. 1985; Gale 1982; Walsh 1981; Gangi 1978). The research has developed from single-fracture to multiple-fractures, and from two-dimensional to three-dimensional, and proposed a variety of empirical formulas and numerical models. Relevant research results will not be discussed in this section. The following will only give a brief summary of the coupling analysis theories of seepage and consolidation in saturated and unsaturated soils.

2.4.1 Consolidation Theories for Saturated Soils

For the consolidation problems of saturated soils, numerous numerical simulation methods have been established. Scholars have studied the seepage and consolidation problems in earth-rock dams from different angles. And the work was mostly conducted based on the Biot's consolidation theory and considered the influence of the coupling effects on the consolidation problems.

Gu (1998) studied the effects of hydraulic conductivity on the consolidation process of earth-rockfill dams under one-way drainage conditions. In the consolidation analysis, the permeability coefficient, k_t, is expressed as a function of the consolidation degree:

$$k_t = k_0(1 - U_t) \tag{2.37}$$

where U_t is the degree of consolidation at time t; and k_0 is the initial hydraulic conductivity of soil before consolidation.

Ping et al. (2001) considered the effects of the stress on the soil permeability in the consolidation analysis of a foundation pit excavation, and expressed the hydraulic conductivity, k, of the soil as a function of water level:

$$k = k_0 \exp(\beta \Delta H) \tag{2.38}$$

where k_0 is the initial hydraulic conductivity; β is an undetermined constant; ΔH is the change of water level.

Chai and Wu (1997) put forward a method of analyzing the interactions between the stress field and the seepage field for consolidation problems of earth-rockfill dams. The hydraulic conductivity of soil is expressed as follows:

$$k = c\frac{e^m}{1 + e} \tag{2.39}$$

where e is the void ration; c and m are undetermined parameters.

Chen et al. (2004) studied the coupling effects of seepage field and stress field on the stability of earth dams and introduced the following relationship between the hydraulic conductivity, k, and the void ratio, e:

$$k = k_0 \exp(\alpha e) \tag{2.40}$$

where k_0 is the initial hydraulic conductivity of soil; α and is a fitting parameter.

Li (2004) proposed a coupling analysis method of stress field and seepage field and expressed the hydraulic conductivity, k, as an exponential function of the effective stress increment, $\Delta\sigma'$:

$$k = k_0 \exp(-\lambda \Delta \sigma') \tag{2.41}$$

where λ is a parameter determined by tests.

Liu et al. (2004) considered the influence of effective stress on the hydraulic conductivity, k, in the seepage and consolidation analysis of a tailing dam. The expression of the hydraulic conductivity is shown as follows:

$$k = k_0 \exp(-\eta \sigma') \tag{2.42}$$

where k_0 is the initial hydraulic conductivity of soil; η is a fitting parameter; σ' is the effective stress.

Lei (2010) analyzed the consolidation of an earth-rockfill dam based on the combination of the Biot's consolidation theory and a hydraulic conductivity model. In the hydraulic conductivity model, the influence of shear stress level and void ratio on the permeability of soil was considered:

$$k = \exp(ae + bS_l + c) \tag{2.43}$$

where e is the void ratio; S_l is the shear stress level; a, b, and c are fitting parameters of the model.

2.4.2 Consolidation Theories for Unsaturated Soils

As for the consolidation problems of unsaturated soils, earlier studies have mainly focused on the air-closed unsaturated soils. The air-containing pore water in the soils was regarded as a single fluid phase. The pore-air phase was ignored in the analysis and only the governing equation of pore-water was established.

The coupled equations of unsaturated consolidation considering the flow of air-phase were first proposed by Barden (1965) in a one-dimensional form. In the equations, the coupling of soil skeleton, pore water, and pore air was taken into account, and the theory of single stress-stress variable was adopted. Fredlund and Hasan (1979) derived the one-dimensional consolidation equations for unsaturated soils based on the theory of double stress-state variables. The flows of pore water and pore air through the soil skeleton were described by the Darcy's law and the Fick's law, respectively. LLoret and Alonso (1980) also studied the one-dimensional consolidation problems in unsaturated soils.

After the 1980s, scholars started to study the multi-dimensional consolidation equations of unsaturated soils. Darkshanamurthy et al. (1984) established three-dimensional coupling equations for unsaturated consolidation analysis based on the equilibrium conditions and the continuity equations of the air-phase and liquid-phase in unsaturated soils. Wong et al. (1998) developed a numerical analysis program for multi-dimensional consolidation problems of unsaturated soils and verified the program with the Mandel-Cryer problem. Oettl et al. (2004) proposed a coupling model for a three-phase medium and analyzed the seepage and consolidation of an earth-rock dam.

Yang (1990) and Yang and Shen (1991, 1992) derived the coupled equations of seepage and consolidation for unsaturated soils based on the theory of double stress-state variables. Considering the influence of temperature on consolidation, the heat conservation equation was incorporated into the coupled equations. And a variety of simplified forms for the coupled equations were also given.

Chen studied the three-dimensional consolidation problems of unsaturated soils based on the theory of mixture in the early 1990s (Chen et al. 1991, 1993; Chen 1993). Thereafter, Chen et al. (2001) extended the constitutive models of saturated soils to unsaturated soils and introduced them into the consolidation theory.

Due to the complexity of the above consolidation equations, they are hard to apply to practical projects for solving complex problems of large-scale. Therefore, many scholars simplified the unsaturated consolidation theories for the purpose of practical application, making it possible to solve complex problems.

Chang and Duncan (1983) studied the consolidation problems of unsaturated soils with a relatively high degree of saturation (greater than 85%) and proposed that the pore water and the pore air can be treated as a mixed-fluid. Based on the above premise, the Biot's consolidation theory was extended to unsaturated soils and formulated for finite element analyses. And a one-dimensional example was analyzed and discussed.

Li et al. (1990) also made an extension of the Biot's consolidation theory for saturated soils to unsaturated soils and analyzed a simple two-dimensional problem by finite element method.

Shen (2003) proposed a simplified consolidation theory for unsaturated soils based on the assumption of pore-air drainage ratio in unsaturated soils. The pore-air continuity equation is no longer included in the governing equations of unsaturated consolidation, which makes the solving difficulty significantly reduced. Then the theory was successfully applied to the numerical simulation of rainfall infiltration of expansive soil slopes (Shen and Mi 2004).

Wei et al. (2006) studied the consolidation problems in soils of high degree of saturation and simplified the pore fluid (i.e., the mixture of pore water and pore air) as a compressible fluid medium. With the simplification, the consolidation equations and their one-dimensional analytic solutions were derived.

Cao and Yin (2009a, b) put forward an improvement for the mixed-fluid simplification method proposed by Chang and Duncan (1983) to solve the problem of unable to calculate the matric suction. The improved method could be used to determine the pore-water pressure, pore-air pressure, and matric suction, and was extended to the study of two-dimensional consolidation problems.

Deng (2005) founded a numerical method for solving the two-dimensional axisymmetric problems based on the assumption proposed by Shen (2003) and applied it to the analysis of clayey soils with surface cracks.

Jie et al. (2012) derived a general form of the finite element equations for the simplified consolidation theory of Shen (2003) and analyzed the influence of deformation on the pore-water pressure in earth-rock dams.

Song (2013) applied the simplified consolidation theory of Shen (2003) to the analysis of two-dimensional engineering examples and studied the effects of the air drainage rate on the pore-air pressure and the rate of consolidation.

2.5 Summary

This chapter summarizes the major research results from aspects of the engineering properties of compacted unsaturated clayey soils, the constitutive models, the strength theories, and the coupling analysis methods of consolidation. Conclusions can be drawn as follows:

(1) In general, there are two kinds of theories for the expression of stress state in unsaturated soils, which are the effective stress theories and the double stress-state variable theories, respectively. Both kinds of theories have advantages as well as disadvantages. And under certain conditions, the two kinds of theories are equivalent. Due to the different expressions of stress state by the two kinds of theories, the strength theories, constitutive models, and consolidation theories of unsaturated soils can also be divided into two categories. Although great progress has been made in unsaturated soil mechanics, there are still many deficiencies and controversies.

(2) For the engineering properties of compacted unsaturated clayey soils, the experimental conditions used to study the coupling relationship between hydraulic and mechanical properties are generally significantly different from the actual conditions in practical engineerings. On the one hand, the studies on the hydraulic and mechanical properties of unsaturated soils are limited by experimental techniques; on the other hand, some important factors are not taken into consideration. For instance, the influence of shearing state has rarely been considered when studying the seepage characteristics.

(3) For the consolidation analysis of earth-rock dams, the multi-physics coupling effects are not adequately considered, and insufficiencies still exist in the analysis methods. The spatial and temporal distribution of the physical state in the core wall is highly inhomogeneous, which will affect the spatial and temporal variation of the soil permeability. Correspondingly, the changes in permeability will then affect the distributions of stress, deformation, and physical-state in the core wall. The current theories and methods still can hardly reflect the complex coupling relationship. Moreover, in practical applications, the consolidation analysis is still based on the framework of the saturated consolidation theory. The unsaturated consolidation theories are still mainly in the stage of theoretical analysis and can hardly be applied to practical problems. Also, the simplified consolidation theories still need further verification and development.

References

Aitchison GD (1961) Relationships of moisture stress and effective stress functions in unsaturated soils. In: Proceedings of conference on pore pressure and suction in soils, London, 1961. Butterworths, pp 47–52

Alonso EE, Cardoso R (2010) Behavior of materials for earth and rockfill dams: Perspective from unsaturated soil mechanics. Front Architect Civil Eng China 4(1):1–39

Alonso EE, Gens A, Josa A (1990) A constitutive model for partially saturated soils. Géotechnique 40(3):405–430

Bao CG (2004) Behavior of unsaturated soil and stability of expansive soil slope. Chin J Geotech Eng 01:1–15

Barden L (1965) Consolidation of compacted and unsaturated clays. Geotechnique 15(3):267–286

Barton N, Bandis S, Bakhtar K (1985) Strength, deformation and conductivity coupling of rock joints. Int J Rock Mech Min Sci Geomech Abs 22(3):121–140

Bian JM, Wang BT (2011) Research on influence of water contents on the shear strength behavior of unsaturated soils. Chin J Undergr Space Eng 7(1):17–21

Bishop AW (1959) The principle of effective stress. Teknisk Ukeblad 39:859–863

Bishop AW, Blight GE (1963) Some aspects of effective stress in saturated and partly saturated soils. Geotechnique 13(3):177–197.

Blight GE (1961) Strength and consolidation characteristics of compacted soils. Doctoral dissertation, Imperial College, London

Bolzon G, Schrefler BA, Zienkiewicz OC (1996) Elastoplastic soil constitutive laws generalized to partially saturated states. Géotechnique 46(2):279–289

Brooks RH, Corey AT (1964) Hydraulic properties of porous media. Colorado State University, Hydrology Papers 3:27

Cao XS, Yin ZZ (2009a) Consolidation method of unsaturated soils for hydraulic fracturing of core walls of rock-fill dams. Chin J Geotech Eng 12:1851–1857

Cao XS, Yin ZZ (2009b) Simplified computation of two-dimensional consolidation of unsaturated soils. Rock Soil Mech 30(9):2575–2580

Carpenter GW, Stephenson RW (1986) Permeability testing in the triaxial cell. Geotech Test J 9(1):3–9

Chai JR (2000) Analysis on coupled seepage and stress fields in dam and its surrounding rock mass. Doctoral dissertation, Xi'an University of Technology, Xi'an

Chai JR, Wu YQ (1997) Research on mathematical model for coupled stress and seepage field in single-zone embankment dam. J Shaanxi Water Power 3:4–7

Chang CS, Duncan JM (1983) Consolidation analysis for partly saturated clay by using an elastic-plastic effective stress–strain model. Int J Numer Anal Meth Geomech 7(1):39–55

Chang ZX, Zhao YS, Hu YQ, Yang D (2004) Theoretic and experimental studies of the coupling of seepage flow and 3D stresses in fractured rock masses. Chin J Rock Mechan Eng 23(S2):4907–4911

Chen M, Chen ZD (1999) Effective stress laws for multi-porosity media. Appl Math Mech 20(11):1121–1127

Chen XP, Qian PY, Liang ZS, Zhang FZ, Wu QX (2004) Coupling analysis of heterogeneous earth dam stability of seepage field and stress field. Rock Soil Mech 25(6):860–864

Chen ZH (1993) Consolidation theory of unsaturated soil based on the theory of mixture (II). Appl Math Mech 14(8):687–698

Chen ZH (1994) Stress state and stress state variables of unsaturated soils. In: Proceeding of the 7th national conference on soil mechanics and foundation engineering, Xi'an, China

Chen ZH (2014) On basic theories of unsaturated soils and special soils. Chin J Geotech Eng 36(2):201–272

Chen ZH, Huang H, Lu ZH (2001) Nonlinear and elasto-plasticity consolidation models of unsaturated soil and applications. Appl Math Mech 22(1):105

Chen ZH, Qin B (2012) On stress state variables of unsaturated soils. Rock Soil Mech 33(1):1–11

Chen ZH, Xie DY, Liu ZD (1991) The consolidation of unsaturated soil. In: Proceedings of seventh international conference on computer methods and advances in geomechanics, Cairns, pp 1617–1621
Chen ZH, Xie DY, Liu ZD (1993) Consolidation theory of unsaturated soil based on the theory of mixture (I). Appl Math Mech 14(2):127–137
Chen ZH, Xie DY, Wang YS (1994) Effective stress in unsaturated soil. Chin J Geotech Eng 03:62–69
Chen ZH, Zhou HQ, Fredlund DG (1998) A non-linear model for unsaturated soils. In: Proceeding of the 2nd international conference on unsaturated soils, Beijing, pp 461–466
Chen ZH, Zhou HQ, Fredlund DG (1999) Nonlinear model for unsaturated soils and its application. Chin J Geotech Eng 5:603–608
Chiu CF, Ng CWW (2003) A state-dependent elasto-plastic model for saturated and unsaturated soils. Géotechnique 53(9):809–829
Chu J, Bo MW, Chang MF, Choa V (2002) Consolidation and permeability properties of Singapore marine clay. J Geotech Geoenviron Eng 128(9):724–732
Coleman JD (1962) Stress/strain relations for partly saturated soils. Geotechnique 12(4):348–350
Darkshanamurthy V, Fredlund DG, Rahardjo H (1984) Coupled three-dimensional consolidation theory of unsaturated porous media. In: Fifth international conference on expansive soils 1984: preprints of papers, Institution of Engineers, Australia, pp 99–103
Deng G (2005) Systematic research on distortion mechanism of over consolidated clay slope. Doctoral dissertation, Tsinghua University, Beijing
Donald IB (1961) The mechanical properties of saturated and partly saturated soils with special reference to negative pore water pressure. Doctoral dissertation, University of London, London
Escario V, Juca J, Coppe MS (1989) Strength and deformation of partly saturated soils. In: Proceedings of the 12th International conference on soil mechanics and foundation engineering, Rio de Janerio, pp 43–46
Fredlund D (1979) Second Canadian geotechnical colloquium: appropriate concepts and technology for unsaturated soils. Can Geotech J 16(1):121–139
Fredlund DG, Hasan JU (1979) One-dimensional consolidation theory: unsaturated soils. Can Geotech J 16(3):521–531
Fredlund DG, Morgenstern NR (1976) Constitutive relations for volume change in unsaturated soils. Can Geotech J 13(3):261–276
Fredlund DG, Morgenstern NR (1977) Stress state variables for unsaturated soils. J Geotech Geoenviron Eng 103 (ASCE 12919)
Fredlund DG, Morgenstern NR, Widger RA (1978) The shear strength of unsaturated soils. Can Geotech J 15(3):313–321
Fredlund DG, Rahardjo H (1993) Soil mechanics for unsaturated soils. Wiley, Canada
Fredlund DG, Xing A, Huang S (1994) Predicting the permeability function for unsaturated soils using the soil-water characteristic curve. Can Geotech J 31(4):533–546
Gale JE (1982) The effects of fracture type (induced versus natural) on the stress-fracture closure-fracture permeability relationships. In: Paper presented at the The 23rd US symposium on rock mechanics (USRMS), Berkeley, California
Gangi AF (1978) Variation of whole and fractured porous rock permeability with confining pressure. Int J Rock Mech Min Sci Geomech Abs 15(5):249–257
Gao YB, Liu FY, Li N (2005) A new method for determination of permeability of unsaturated soils. Chin J Rock Mechan Eng 24(18):3258–3261
Gardner WR (1958) Some steady-state solutions of the unsaturated moisture flow equation with application to evaporation from a water table. Soil Sci 85(4):228–232
Gu WC (1998) Consolidation analysis of earth-rock dam considering the variation of permeability coefficient. Hebei Water Resour Hydropower Eng 4:13–17
Houlsby GT (1997) The work input to an unsaturated granular material. Géotechnique 47(1):193–196
Hu B, Wang R, Wang Z, Gong BW (2008) Research on comparison of strength equations of unsaturated soils. Chin J Rock Mechan Eng 27(S1):3005–3009

Huang H, Chen ZH, Li G (2000) A study on yield locus of unsaturated soils on *p-s* plane and soil-water characteristic curve. Rock Soil Mech 21(4):316–321

Huang RQ, Wu LZ (2007) Study on the shear strength of unsaturated expansive soils. J Chengdu Univ Technol (Science & Technology Edition) 34 (3):221–224

Huang S, Barbour SL, Fredlund DG (1998) Development and verification of a coefficient of permeability function for a deformable unsaturated soil. Can Geotech J 35(3):411–425

Jennings JE (1961) A revised effective stress law for use in the prediction of the behaviour of unsaturated soils. In: Proceedings of conference on pore pressure and suction in soils, London, Butterworth, pp 26–30

Jie YX, Wen YF, Deng G, Chen R, Xu ZP (2012) Impact of soil deformation on phreatic line in earth-fill dams. Comput Geosci 46:44–50

Khalili N, Khabbaz MH (1995) On the theory of three-dimensional consolidation in unsaturated soils. In: Proceedings of the first international conference on unsaturated soils (UNSAT 95), Paris, France

Khalili N, Khabbaz MH (1998) A unique relationship for χ for the determination of the shear strength of unsaturated soils. Geotechnique 48(5):681–687

Kohgo Y, Nakano M, Miyazaki T (1993) Theoretical aspects of constitutive modelling for unsaturated soils. Soils Found 33(4):49–63

Lei HJ (2010) A study on seepage characteristics of clayey soil of high earth-rockfill dam with large shear deformation. Doctoral dissertation, Tsinghua University, Beijing

Lei HJ, Wu YK, Yu YZ, Zhang BY, Lv H (2016) Influence of shear on permeability of clayey soil. Int J Geomech 16(5):04016010

Leong EC, Rahardjo H (1997) Permeability functions for unsaturated soils. J Geotech Geoenviron Eng 123(12):1118–1126

Li XK, Zienkiewicz OC, Xie YM (1990) A numerical model for immiscible two-phase fluid flow in a porous medium and its time domain solution. Int J Numer Meth Eng 30(6):1195–1212

Li XY (2004) Fully coupling analysis of seepage and stress for deep foundation pit based on displacement back analysis. Chin J Rock Mechan Eng 8:1269–1274

Li YL, Liu CR, Liu HN, Liu HQ (2004) Testing study on permeability characteristics of unsaturated soil. Chin J Rock Mechan Eng 23(22):3862–3865

Ling H, Yin ZZ (2007) Variation of unsaturated soil strength with water contents. Chin J Rock Mechan Eng 26(7):1499–1503

Ling H, Yin ZZ, Cai ZY (2008) Experimental study on stress-water content-strain relationship of unsaturated soil. Rock Soil Mech 29(3):651–655

Liu CH, Chen CX, Fu SL (2002a) Testing study on seepage characteristic of a single rock fracture under two-dimensional stresses. Chin J Rock Mechan Eng 21(8):1194–1198

Liu CH, Chen CX, Fu SL (2002b) Testing study on seepage characteristics of single fracture with sand under shearing displacement. Chin J Rock Mechan Eng 21(10):1457–1461

Liu HL, Li N, Liao X, Gong JL, Fang FH (2004) Unsteady seepage analysis of tailings dams considering coupling of stress and seepage fields. Chin J Rock Mechan Eng 23(17):2870–2875

Liu J (1992) Seepage stability and seepage control of soil. China Water Power Press, Beijing

LLoret A, Alonso EE (1980) Consolidation of unsaturated soils including swelling and collapse behaviour. Géotechnique 30(4):449–477

Lu N (2008) Is matric suction a stress variable? J Geotech Geoenviron Eng 134(7):899–905

Marshall TJ (1958) A relation between permeability and size distribution of pores. Eur J Soil Sci 9(1):1–8

McKee CR, Bumb AC (1987) Flow-testing coalbed methane production wells in the presence of water and gas. SPE Formation Eval 2(04):599–608

Mctigue DF, Wilson RK, Nunziato JW (1983) An effective stress principle for partially saturated granular media. Stud Appl Mech 7:195–210

Mesri G, Olson RE (1971) Mechanisms controlling the permeability of clays. Clays Clay Miner 19(3):151–158

Miao LC (2007) Research of constitutive model of unsaturated soils. Rock Soil Mech 28(5):855–860

Miller CJ, Yesiller N, Yaldo K, Merayyan S (2002) Impact of soil type and compaction conditions on soil water characteristic. J Geotech Geoenviron Eng 128(9):733–742

Modaressi A, Abou-Bekr N (1994) A unified approach to model the behaviour of saturated and unsaturated soils. In: Proceedings of the 8th international conference on computer methods and advances in geomechanics, pp 1507–1513

Nikolaevsky VN (1985) Mechanics of fluid-saturated geomaterials. Mech Geomat 379–399

Öberg AL, Sällfors G (1997) Determination of shear strength parameters of unsaturated silts and sands based on the water retention curve. Geotech Test J 20(1):40–48

Oettl G, Stark RF, Hofstetter G (2004) Numerical simulation of geotechnical problems based on a multi-phase finite element approach. Comput Geotech 31(8):643–664

Pereira JHF, Fredlund DG (2000) Volume change behavior of collapsible compacted gneiss soil. J Geotech Geoenviron Eng 126(10):907–916

Ping Y, Bai SW, Xu YP (2001) Numerical simulation of seepage and stress coupling analysis in deep foundation pit. Rock Soil Mech 22(1):37–41

Samarasinghe AM, Huang YH, Drnevich VP (1982) Permeability and consolidation of normally consolidated soils. J Geotech Eng Div 108(6):835–850

Shen XZ, Guan XJ, Lan Y (2007) Calculation of effective strength indexes of unsaturated low liquid limit clay. Rock Soil Mech 28(S1):207–210

Shen ZJ (1995) Reduced suction and simplified consolidation theory for expansive soils. In: Proceedings of the first international conference of unsaturated soils (UNSAT 95), Paris, France

Shen ZJ (2000) Theoretical soil mechanics. China Water & Power Press, Beijing

Shen ZJ (2003) Simplified consolidation theory for unsaturated soils and its application. Hydroscience Eng 4:1–6

Shen ZJ, Mi ZK (2004) Coupled analyses of seepage and deformation of expansive soil slopes during rainfall. Hydro-Science Eng 3:7–11

Song Y (2013) Simplified consolidation theory for unsaturated soils and its application in civil engineering. M.S. thesis, Beijing Jiaotong University, Beijing

Su BY, Zhan ML, Wang Y (1994) Experimental study on the coupling characteristics of fracture seepage and stresses. Chin J Geotech Eng 19(4):73–77

Sun DA, Sheng D, Sloan SW (2007) Elastoplastic modelling of hydraulic and stress–strain behaviour of unsaturated soils. Mech Mater 39(3):212–221

van Genuchten MT (1980) A closed-form equation for predicting the hydraulic conductivity of unsaturated soils. Soil Sci Soc Am J 44(5):892–898

Vanapalli SK, Fredlund DG, Pufahl DE (1999) The influence of soil structure and stress history on the soil-water characteristics of a compacted till. Géotechnique 49(2):143–159. https://doi.org/10.1680/geot.1999.49.2.143

Vanapalli SK, Fredlund DG, Pufahl DE, Clifton AW (1996) Model for the prediction of shear strength with respect to soil suction. Can Geotech J 33(3):379–392

Vardoulakis I, Beskos DE (1986) Dynamic behavior of nearly saturated porous media. Mech Mater 5(1):87–108

Walsh JB (1981) Effect of pore pressure and confining pressure on fracture permeability. Int J Rock Mech Mining Sci Geomech Abs 18(5):429–435

Wang DL, Luan MT, Yang Q (2009) Experimental study of soil-water characteristic curve of remolded unsaturated clay. Rock Soil Mech 30(3):751–756

Wang Y, Xu ZY, Su BY (2000) Complete coupled analysis of fluid flow and elastoplastic stress in complicated fractured rock masses. Chin J Rock Mechan Eng 19(2):177–181

Wei HY, Zhan LT, Chen YM (2006) Compressibility and consolidation of unsaturated soils with high degree of saturation and its application. Chin J Geotech Eng 28(2):264–269

Wheeler SJ, Sharma RS, Buisson MSR (2003) Coupling of hydraulic hysteresis and stress–strain behaviour in unsaturated soils. Géotechnique 53(1):41–54

Williams J, Prebble RE, Williams WT, Hignett CT (1983) The influence of texture, structure and clay mineralogy on the soil moisture characteristic. Soil Res 21(1):15–32

Wong TT, Fredlund DG, Krahn J (1998) A numerical study of coupled consolidation in unsaturated soils. Can Geotech J 35(6):926–937

Yang DQ (1990) Generalized unsaturated soil consolidation theory and numerical simulation and experimental study. Doctoral dissertation, Nanjing Hydraulic Research Institute, Nanjing, China

Yang DQ, Shen ZJ (1991) Two-dimensional numerical simulation of generalized consolidation problem of unsaturated soils. In: Proceedings of the seventh international conference on computer methods and advances in geomechanics, Cairns

Yang DQ, Shen ZJ (1992) Study on generalized consolidation theory of unsaturated soils. In: Proceedings of the 7th international conference on expansive soils, Dallas

Ye YX, Liu GT (2005) Research on coupling characteristics of fluid flow and stress within rock. Chin J Rock Mechan Eng 24(14):2518–2525

Yin JH (2009) Influence of relative compaction on the hydraulic conductivity of completely decomposed granite in Hong Kong. Can Geotech J 46(10):1229–1235

Yin ZZ, Zhou J, Chiu CF, Yuan JP, Zhang KY (2006) Constitutive relations and deformation calculation for unsaturated soils. Chin J Geotech Eng 28(2):137–146

Yu PJ, Chen YJ (1965) The pore air-water configurations and their effects on the mechanical properties of partially saturated soils. J Hydraul Eng 01:16–24

Zhang XD, Zhao CG, Liu Y (2010) Probability based model for influence of deformation on hydraulic conductivity function of unsaturated soils. J Eng Geol 18(1):132–139

Zhao CG, Liu Y (2009) Continuum porous medium soil mechanics and its application in constitutive relationship of unsaturated soils. Chin J Geotech Eng 9:1324–1335

Zhao CG, Liu Y, Gao FP (2010) Work and energy equations and the principle of generalized effective stress for unsaturated soils. Int J Numer Anal Meth Geomech 34(9):920–936

Chapter 3
Study on the Mechanical Properties of Clayey Core Material

The stress state of an unsaturated soil is different from that of the saturated soils under the same boundary and load conditions. The matric suction in unsaturated soils plays an important role in the soil's stress state. And thus, the strength and modulus of the unsaturated soil are also significantly different from those of the saturated soil. Moreover, the stress state and the degree of saturation of the core material in the dam will change gradually during construction and impounding, which brings significant impacts on the stress and deformation of the dam. In this chapter, both saturated and unsaturated conventional triaxial consolidated-drained tests are carried out using the core material of gravelly clay from the Nuozhadu high earth-rockfill dam to reveal the strength and deformation characteristics of the core material in different saturation states. Then, on the basis of the test results, the strength and modulus differences between the unsaturated soil and the saturated soil are analyzed.

3.1 Test Material

The core material of the Nuozhadu high earth-rockfill dam was used as the test material in this chapter. The Nuozhadu dam, with a height of 261.5 m, is the highest earth-rock dam in China. During the construction of the dam, a certain amount of gravel was mixed into the clay for use as impervious core material to improve the stress-deformation behavior of the core wall. The gravelly clay was prepared by adding gravels of fresh breccia and granite into the natural clay material, with a gravel ratio of 35%. Figure 3.1 shows the particle size grading curve of the core material.

In the following triaxial tests, in order to simulate the actual state of the core material in the dam after roller compaction, the particle size grading, the dry density, the water content, and other physical parameters were designed on the basis of the actual construction situation of the dam.

Y. Wu, *Multi-physics Coupling Analysis of Clayey Core Wall of High Earth-Rockfill Dam*, Springer Theses, https://doi.org/10.1007/978-981-13-1032-4_3

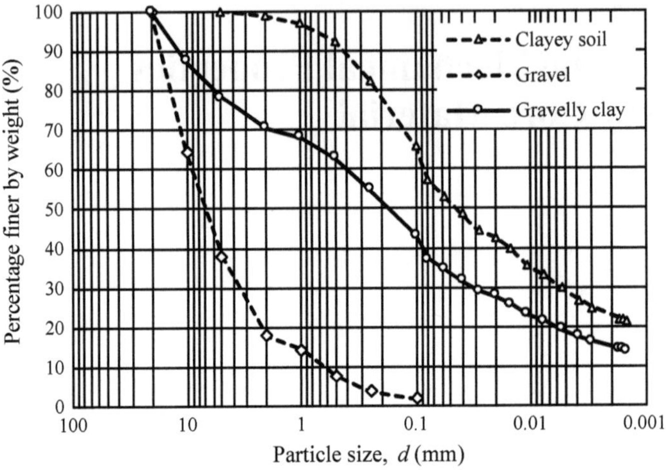

Fig. 3.1 Particle size grading curve of the Nuozhadu core material

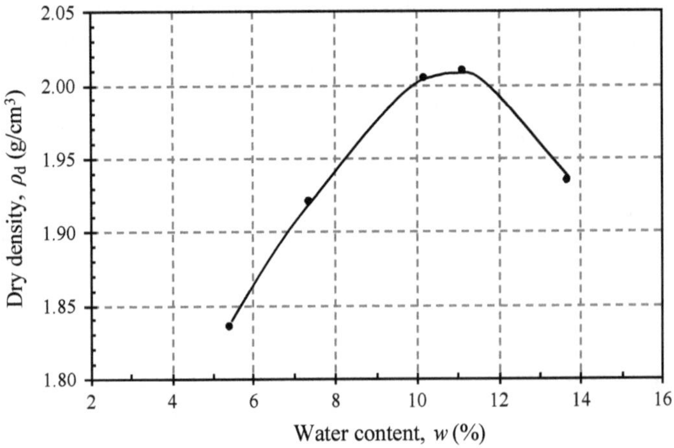

Fig. 3.2 Compaction curve of the Nuozhadu core material

Then compaction tests were performed using the Nuozhadu core material to determine the optimum water content of the soil. Based on the test results shown in Fig. 3.2, the optimum water content, w_{opt}, of the soil would be 11.1%.

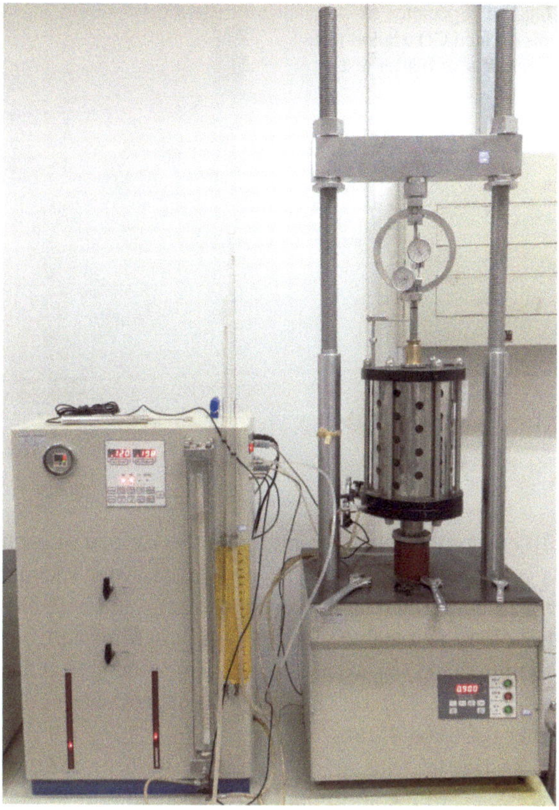

Fig. 3.3 Strain controlled triaxial apparatus

3.2 Apparatus and Method for Triaxial Testing

3.2.1 Testing Apparatus

The conventional triaxial apparatus at the Institute of Geotechnical Engineering at Tsinghua University, as shown in Fig. 3.3, was used to conduct the tests. The maximum confining pressure that can be applied in the triaxial tests is 2000 kPa.

3.2.2 Specimen Preparation

The specimens for the triaxial tests were prepared with the Nuozhadu core material with the water content of 11.1% and the dry density of 1.96 g/cm^3. The specimens were cylinder-shaped, with a height of 20 cm and a diameter of 10 cm. The specimen was compacted in 5 layers by manual compaction, each having a thickness of 20 mm.

Fig. 3.4 Specimen for unsaturated CD triaxial test: **a** before test; **b** after test

To ensure the specimen integrity, the surface of each layer was roughened to form a connection zone with the next layer. After specimen preparation, the specimens for saturated triaxial tests were vacuum saturated, and the specimens for unsaturated triaxial tests were wrapped with plastic films and stored in sealing bags. Thereafter, saturated and unsaturated consolidated-drained (CD) triaxial tests were conducted, respectively. Figure 3.4 shows the appearance of a specimen after preparation and after the triaxial test.

During the triaxial tests, the axial strain, ε_1, the axial stress, σ_1, and the volumetric strain, ε_v, of the specimen were recorded. For the saturated CD tests, the specimen shrinkage could be obtained by measuring the volume of the drained water. For unsaturated CD tests, due to the unsaturated state of the specimen, pore air as well as pore water (if there is any) will drain out of the specimen. Hence the volumetric strain cannot be obtained directly by measuring the pore-water drainage. In this dissertation, the volumetric strain of the specimen for unsaturated tests was obtained indirectly through measuring the volume change of the chamber fluid and the volume of the axial load piston entering into the chamber (with correction of the chamber deformation).

3.2.3 Experimental Scheme

Two sets of conventional triaxial CD tests were performed under different confining pressures using the clayey core wall material of the Nuozhadu High earth-rockfill dam. The first set of tests was conducted using the saturated specimens and the second set of tests using the unsaturated specimens. Table 3.1 gives the detailed experimental scheme.

Table 3.1 Scheme of the triaxial tests

Type	Confining pressure (kPa)					
Saturated CD test	100	200	400	800	1300	2000
Unsaturated CD test	100	200	400	800	1300	2000

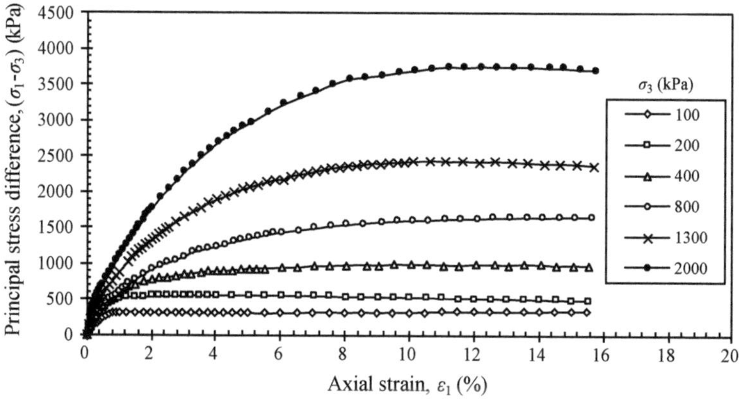

Fig. 3.5 Relationship between the principal stress difference, $(\sigma_1 - \sigma_3)$, and the axial strain, ε_1, for saturated tests

3.3 Results of the Triaxial Tests

3.3.1 *Saturated Test Results*

Figures 3.5 and 3.6 show the curves of the principal stress difference, $(\sigma_1 - \sigma_3)$, versus the axial strain, ε_1, and the curves of the volumetric strain, ε_v, versus the axial strain, ε_1, respectively. It can be seen from Fig. 3.5 that under different confining pressures, the principal stress difference increases with the increase of the axial strain and gradually trends to a stable value. The stress-strain relationship is of strain hardening. Comparing the test curves of different confining pressures, as the confining pressure increases, the maximum and ultimate values of the principal stress difference continuously increase.

The confining pressure is not only related to the peak strength of soil but also has significant influence on the stress-strain relationship and the volumetric strain. In Fig. 3.6, for lower confining pressures (e.g., 100 kPa), the specimen is slightly dilated with the increase of the axial strain. For higher confining pressures, the specimen shrinks as it is sheared and the volume shrinkage increases with the increase of the axial strain. Comparing the curves at confining pressures of 100, 200, 400, and 800 kPa, the volumetric strain varies significantly. When the confining pressure is greater than 800 kPa, the difference between the volumetric strain curves apparently reduces.

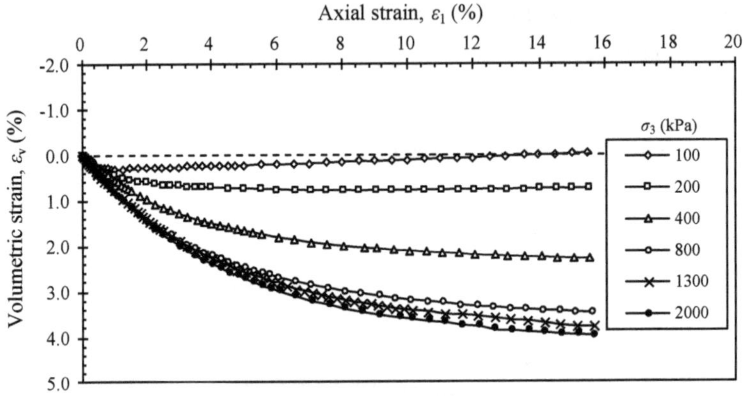

Fig. 3.6 Relationship between the volumetric strain, ε_v, and the axial strain, ε_1, for saturated tests

Fig. 3.7 Relationship between the void ratio, *e*, and the confining pressure, σ_3

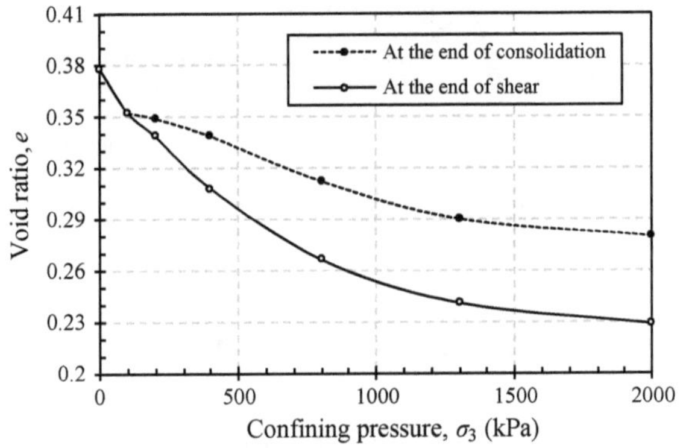

Figure 3.7 shows the relationship between the void ratio, *e*, and the confining pressure, σ_3, of the triaxial tests. The dashed line and the solid line in the figure give the void ratio of the specimens at the end of consolidation stage and after shearing, respectively. The initial void ratio of the specimens is about 0.38. With the increasing of the confining pressure, the change rate of the specimen volume decreases. The reason is that with the decrease of the void ratio, the clay particles are squeezed close to each other, and the interparticle force increases, thereby increasing the difficulty of compression.

The Mohr circle for the saturated triaxial CD tests is shown in Fig. 3.8. The strength index can be determined as $c = 63.8$ kPa, and $\varphi = 27.9°$. Moreover, with the increase of the confining pressure, the failure envelope shows a tendency to curve downward.

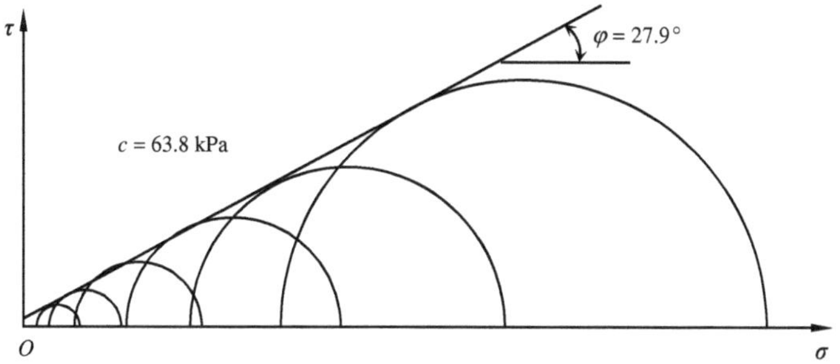

Fig. 3.8 Mohr circles and the failure envelope of the saturated CD triaxial tests

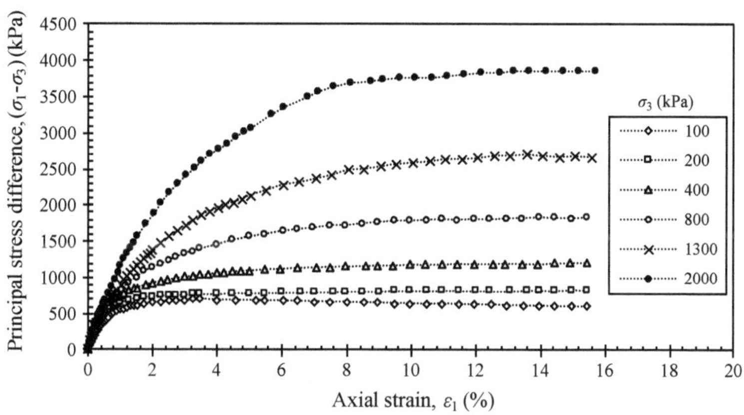

Fig. 3.9 Relationship between the principal stress difference, $(\sigma_1 - \sigma_3)$, and the axial strain, ε_1, for unsaturated tests

3.3.2 *Unsaturated Test Results*

Figures 3.9 and 3.10 show the curves of the principal stress difference, $(\sigma_1 - \sigma_3)$, versus the axial strain, ε_1, and the curves of the volumetric strain, ε_v, versus the axial strain, ε_1, for the unsaturated triaxial tests, respectively. The distributions of the curves in these figures are similar to those of the saturated test results, and the stress-strain relationship is also of strain hardening.

In the above unsaturated tests, no pore water was drained out from the specimen. Therefore, this set of tests can also be regarded as the constant water content tests. Although the water content of the specimen remains constant during the test, the saturation and matric suction of the specimen continuously vary due to the changing of the specimen volume and porosity. During the unsaturated tests, except for the tests under low confining pressures (i.e., 100 and 200 kPa), the degree of saturation of the specimen gradually increases, and correspondingly, the matric suction decreases.

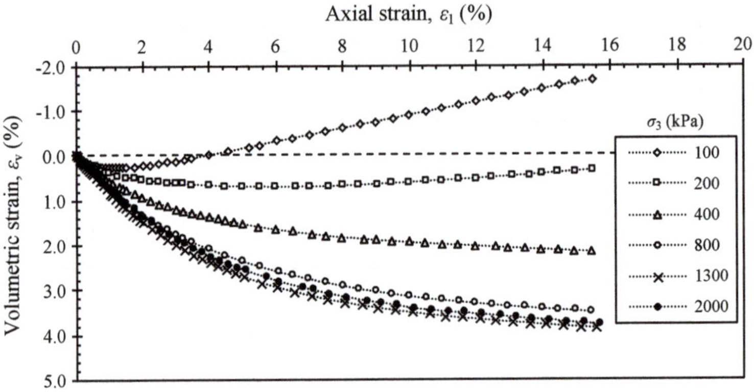

Fig. 3.10 Relationship between the volumetric strain, ε_v, and the axial strain, ε_1, for unsaturated tests

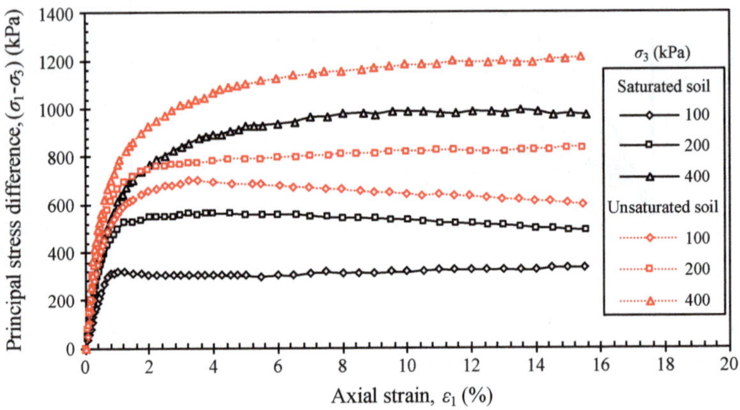

Fig. 3.11 Comparison of the $(\sigma_1 - \sigma_3)$–ε_1 curves for saturated and unsaturated tests

3.3.3 Comparison of the Test Results

The comparison of the results between the two sets of CD triaxial tests is shown in Figs. 3.11, 3.12, 3.13 and 3.14, in which the solid lines represent the data of the saturated triaxial tests, and the dashed lines represent the data of the unsaturated triaxial tests. As shown in the figures, under relatively low confining pressures (e.g., 100, 200, and 400 kPa), both the strength and bulk modulus of the unsaturated soil are apparently higher than that of the saturated soil. While under relatively high confining pressures (e.g., 800, 1300, and 2000 kPa), the difference between the saturated and unsaturated test results significantly decreases.

For the unsaturated triaxial tests, the matric suction in the unsaturated soil specimens increases the effective stress, which significantly affects the strength and defor-

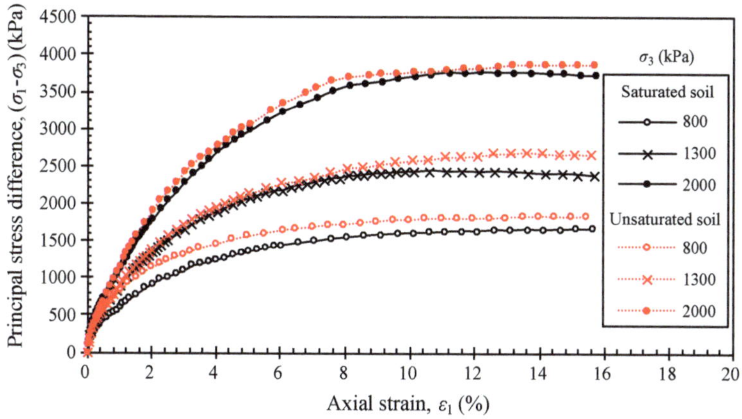

Fig. 3.12 Comparison of the $(\sigma_1 - \sigma_3)$–ε_1 curves for saturated and unsaturated tests

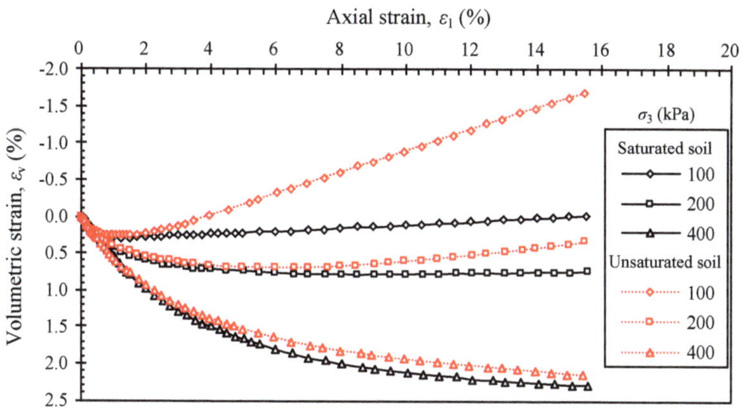

Fig. 3.13 Comparison of the ε_v–ε_1 curves for saturated and unsaturated tests

mation characteristics of the soil. Although the magnitude of the matric suction is continually changing during the tests, its positive contribution to the effective stress always exists.

For the tests of low confining pressures, the degree of saturation of the unsaturated specimen is relatively low due to smaller shrinkage during test. And the influence of the matric suction on the effective stress will be significant, which makes that the difference between the curves of the saturated and unsaturated tests is enormous. With the increase of the confining pressure, the volumetric shrinkage of the unsaturated specimen increases, and its degree of saturation increases. Accordingly, the influence of the matric suction on the effective stress reduces. Hence, the curves of the unsaturated tests gradually approach the curves of the saturated samples (Figs. 3.11 and 3.12).

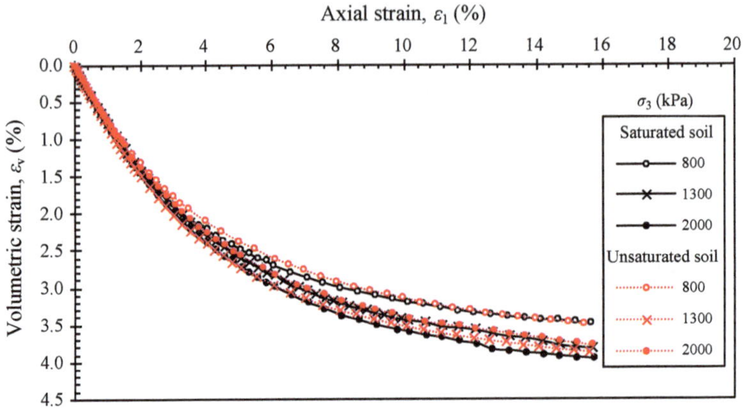

Fig. 3.14 Comparison of the ε_v–ε_1 curves for saturated and unsaturated tests

The stress path for the unsaturated triaxial CD tests of this chapter can be illustrated by Fig. 3.15. Stress point A represents the initial stress state of the soil at the end of consolidation under the net confining pressure, $(\sigma_3 - u_a)$, and the matric suction, $(u_a - u_w)$. In the shearing stage of the test, the confining pressure, $(\sigma_3 - u_a)$, remains constant, and the axial stress, $(\sigma_1 - u_a)$, increases continuously. If the specimen shrinks during shear, its degree of saturation will increase, and accordingly, the matric suction will decrease. The Mohr circle will continuously move along the negative direction of the matric suction coordinate axis with a constant minor principal stress, $(\sigma_3 - u_a)$, until it reaches the failure envelope. In this process, the stress state of the soil specimen gradually moves from stress point A to stress point B. Stress point B represents the stress state at failure, and the failure envelope is tangent to the Mohr circle at stress point C. Path AB will be the stress path for the unsaturated CD test (also can be regarded as a constant water content test).

3.4 Summary

In this chapter, two sets of saturated and unsaturated conventional triaxial consolidated-drained tests were carried out using the core material of gravelly clay from the Nuozhadu high earth-rockfill dam, and the mechanical properties of the gravelly clay material were analyzed. To simulate the actual state of the core material in the dam, the soil gradation, the dry density, the water content, etc., of the test specimens were designed according to the actual construction situation of the core material. The main conclusions can be drawn as follows:

(1) Comparing the experimental results with different confining pressures, the initial tangent modulus, the bulk modulus, the strength, as well as the shrinkage of the specimens increase with the increase of the confining pressure. As the volume

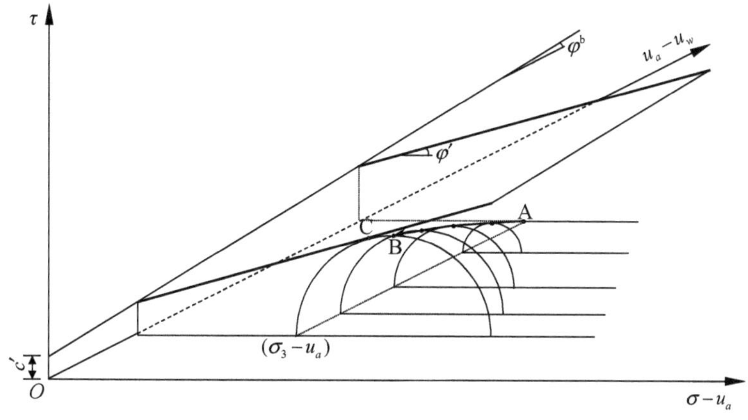

Fig. 3.15 Stress path of the unsaturated CD (constant water content) test

(or the void ratio) of the specimen decreases, the clay particles move closer to each other, and the interparticle force increases gradually, thereby increasing the difficulty of further compression.

(2) The mechanical properties of unsaturated soil are significantly different from those of saturated soil. With the same experimental conditions, the strength and modulus of the unsaturated soil are higher than those of the saturated soil, especially for the results of low confining pressures. When the soil is unsaturated, the matric suction in the soil plays an important role in the effective stress, which will significantly affect the strength and deformation properties of the soil. Moreover, the results also indicate that the mechanical properties of the core material of the dam will change significantly in the process of construction and impounding.

(3) In the unsaturated triaxial tests, with the increase of the soil specimen shrinkage, the degree of saturation of the specimen increases. For high confining pressures, the soil specimen gradually approaches or even reaches the state of completely saturated during the test, and meanwhile, the effect of matric suction on the effective stress gradually weakens. Therefore, with the increase of confining pressure, the difference of mechanical properties between the unsaturated soil and the saturated soil tends to decrease.

(4) For the reason that the mechanical properties of the core material will change obviously with the degree of saturation during construction and impounding, the role of the matric suction (or the degree of saturation) should be fully considered in the stress-deformation analysis of unsaturated soils.

Chapter 4
Study on the Consolidation Theories for Unsaturated Soils

A brief overview of the development of the consolidation theories for unsaturated soils is given in Chap. 2. The theories still involve many limitations and difficulties and cannot be applied to solve large-scale problems in practical engineerings. This is the main reason why the saturated consolidation theories are generally adopted instead of the unsaturated consolidation theories in the consolidation analysis of engineering applications. Moreover, the multi-physics coupling effects, which are particularly significant in high earth-rockfill dams, have rarely been taken into consideration in the current consolidation analyses of earth-rock dams. In this chapter, a simplified consolidation theory for unsaturated soils proposed by Shen (2003) is introduced. Based on the simplified theory, the governing equations of seepage-consolidation for saturated-unsaturated soils are derived, and then a more comprehensive and reasonable multi-physics coupling analysis method is proposed to describe the interactions between different fields. Finally, core issues for the coupling analysis method (e.g., the selection of constitutive models, the description of soil permeability, and the calculation of effective stress) are clarified.

4.1 Theory Basis and Basic Assumptions

4.1.1 Stress-State Variables and the Unsaturated Consolidation Theory

Stress state variables are those that are required to completely describe the stress state of a soil. The selection of the stress-state variables is the most basic and important issue in the consolidation analysis of unsaturated soils. In the unsaturated soil mechanics, there are mainly two ways to describe the stress state of unsaturated soils, i.e., the single stress-state variable theories and the double stress-state variable theories.

Y. Wu, *Multi-physics Coupling Analysis of Clayey Core Wall of High Earth-Rockfill Dam*, Springer Theses, https://doi.org/10.1007/978-981-13-1032-4_4

The theory of double stress-state variables was firstly proposed by Coleman (1962). He suggested that the stress-state variables, $(\sigma - u_a)$ and $(u_a - u_w)$, can be used to describe the stress state of unsaturated soils while ignoring the compression of soil particles and pore water. This concept has been widely accepted and studied. However, the consolidation theories established based on the double stress-state variables are very complicated to use in practical applications and cannot draw on the mature results in saturated soil mechanics. And it is necessary to reestablish the theories of strength, deformation, as well as consolidation for unsaturated soils.

For saturated soils, the single stress-state variable of effective stress has been successfully applied. In view of this, numerous attempts have been made to develop a similar concept of effective stress for unsaturated soils. At present, scholars have generally believed that for non-collapsible soils under certain conditions, both the concept of single stress-state variable and the concept of double stress-state variables can be used to describe the stress state of unsaturated soils. To some extent, the two concepts are equivalent. For unsaturated soils, the expression of the stress state using single stress-state variable is very simple and similar to the expression of stress state in saturated soils. Therefore, related results and ideas in the saturated soil mechanics can be introduced into the studies of unsaturated soils, which significantly improves the practicality of the single stress-state variable theory.

Scholars have also made numerous attempts to study the consolidation theory of unsaturated soils. Based on the theory of the double stress-state variables, different types of coupling equations of consolidation for unsaturated soil, including the equilibrium equations and the continuity equations of pore water and pore air, have been derived from different angles in the 1990s. In these equations, there are a total of five unknown variables, namely three unknown displacement variables, an unknown pore-water pressure variable, and an unknown pore-air pressure variable. Compared with the coupling equations with four unknown variables for saturated soils, although the above theories only have one more unknown variable, the difficulty of solving the equations is greatly increased. Except for few theoretical solutions obtained in one-dimensional conditions, the consolidation theories of unsaturated soils with two stress-state variables has rarely been used in engineering applications. The practicality of the theories has been greatly restricted.

In order to promote the practical application of consolidation theories for unsaturated soils, Shen (2003) proposed a simplified consolidation theory for unsaturated soils on the basis of the single stress-state variable theory in 2003. Then Deng (2005) and Jie et al. (2012) carried out further studies based on the simplified theory and made further improvements. The core idea of the simplified consolidation theory is to simplify the pore-air phase in unsaturated soils. By this means, the pore-air continuity equation will be removed from the consolidation coupling equations, making it possible to solve complex unsaturated consolidation problems.

In this dissertation, the works on the multi-physics coupling consolidation theory of unsaturated soils are carried out based on the simplified theory proposed by Shen. The simplified theory adopts the single-stress state variable. And the pore-air phase in the unsaturated soil is simplified, which significantly reduces the difficulty of solving the unsaturated consolidation problems. On the basis of the previous studies, a more reasonable multi-physics coupling analysis method is developed, which makes it possible to perform multi-physics coupling consolidation analysis of unsaturated soils.

4.1.2 Basic Assumptions for the Simplified Consolidation Theory

The research work in this dissertation is performed based on the simplified consolidation theory and focuses on the core issues in the seepage-consolidation analysis of unsaturated soils. The main simplifications and assumptions are as follows:

(1) The draining of pore air is controlled by the parameter of the air drainage ratio;
(2) Only small deformation occurs in soil skeleton;
(3) The soil particles and pore water are incompressible;
(4) The effects of temperature are ignored;
(5) The flow of pore air is ignored, and the pore-air pressure is calculated by the simplified method;
(6) The movement of the air dissolved in the pore water is ignored;
(7) The water vapor in the pore air is ignored;
(8) The pore water in the soil is interconnected;
(9) The flow of pore water obeys Darcy's law;
(10) The compression of pore air obeys Boyle's law;
(11) The dissolution of pore air in pore water obeys Henry's law.

4.2 Calculation of the Pore-Air Pressure in the Simplified Consolidation Theory

The simplification method proposed by Shen (2003) is introduced in this chapter. The core idea of this method is to simplify the calculation of the pore-air phase so that the continuity equation of pore-air will be omitted from the coupling equations of consolidation. And the difficulty in solving unsaturated consolidation problems could be greatly reduced.

In the simplified consolidation theory, Shen (2003, 2006) defined the concept of the pore-air content, which is the pore-air volume (including the volume of the air dissolved in the pore water) in unit volume of soil. The pore-air content can be expressed as:

$$n_a = [1 - (1 - c_h)S_r]n \tag{4.1}$$

where n is the soil porosity; S_r is the degree of saturation; c_h is the volumetric coefficient of air solubility, which is approximately equal to 0.02 at the temperature of 20 °C.

For unsaturated seepage-consolidation analysis, the continuity of the pore air in unsaturated soils can be described by the following equation:

$$\frac{\partial}{\partial x}\left(\rho_a k_{ax}\frac{\partial u_a}{\partial x}\right)+\frac{\partial}{\partial y}\left(\rho_a k_{ay}\frac{\partial u_a}{\partial y}\right)+\frac{\partial}{\partial z}\left(\rho_a k_{az}\frac{\partial u_a}{\partial z}\right)=\frac{\partial}{\partial t}(\rho_a(1-S_r)n+\rho_a c_h S_r n) \tag{4.2}$$

where u_a is the pore-water pressure; ρ_a is the density of the pore air; k_a is the air coefficient of permeability (also known as the hydraulic conductivity of air).

Considering the definition of the pore-air content in Eq. (4.1), the continuity equation of pore air presented in Eq. (4.2) can be further written as follows:

$$\frac{\partial}{\partial x}\left(\rho_a k_{ax}\frac{\partial u_a}{\partial x}\right)+\frac{\partial}{\partial y}\left(\rho_a k_{ay}\frac{\partial u_a}{\partial y}\right)+\frac{\partial}{\partial z}\left(\rho_a k_{az}\frac{\partial u_a}{\partial z}\right)=\frac{\partial}{\partial t}(\rho_a n_a) \tag{4.3}$$

For almost all the practical problems, the initial state of pore-air pressure can be set to 0 (where the pore-air pressure is the relative pressure; its absolute value equals to the value of the atmospheric pressure, p_a), and the initial pore-air content is denoted as n_{a0}. When the soil is completely undrained, the mass of the pore air in the soil will keep constant. According to Boyle's law, the pore-air pressure is proportional to the density of pore air:

$$\frac{\rho_a}{\rho_{a0}}=\frac{p_a+u_a}{p_a} \tag{4.4}$$

where ρ_{a0} is the density of pore air at atmospheric pressure, p_a.

Hence, when the pore-air pressure is u_a, the density of pore air can be expressed as:

$$\rho_a=\rho_{a0}\left(1+\frac{u_a}{p_a}\right) \tag{4.5}$$

When the soil is completely undrained, both the left side and the right side of the continuity equation of pore air are zero. Then at any time t, there is $\rho_a n_a=\rho_{a0}n_{a0}$. Considering that the density of pore air in this equation can be calculated by Eq. (4.5), the following equation can be obtained:

$$\rho_{a0}\left(1+\frac{u_a}{p_a}\right)\cdot n_a=\rho_{a0}n_{a0} \tag{4.6}$$

By rearranging Eq. (4.6), the pore-air pressure, while the soil is completely undrained, can be further obtained:

$$u_a = \left(\frac{n_{a0}}{n_a} - 1\right)p_a \tag{4.7}$$

where n_{a0} is the initial pore-air content, and $n_{a0} = [1 - (1 - c_h)S_{r0}]n_0$; n_0 is the initial soil porosity; S_{r0} is the initial degree of saturation.

For general conditions that the pore air is partially drained, the air drainage ratio can be defined as

$$\xi = \frac{\Delta q_a}{\rho_a \Delta n_a} \tag{4.8}$$

where Δq_a is the mass of the drained pore air within a period of Δt; $\rho_a \Delta n_a$ is the total mass of pore air that can be drained from the soil. The air drainage ratio, ξ, is one of the most important concepts in this simplified consolidation theory for unsaturated soil.

Based on the continuity equation of pore air, the following equation can be obtained:

$$\Delta q_a = \Delta(\rho_a n_a) \tag{4.9}$$

Δq_a can be further expressed as follows:

$$\Delta q_a = n_a \Delta \rho_a + \rho_a \Delta n_a \tag{4.10}$$

According to the ideal gas law, i.e., $pV = nRT$, the following equation can be obtained:

$$\frac{p_1 V_1}{m_1} = \frac{p_2 V_2}{m_2} = \text{constant} \Rightarrow \frac{p_1}{\rho_1} = \frac{p_2}{\rho_2} = \text{constant} \tag{4.11}$$

Then, the relationship between the pore-air density, ρ, and the pore-air pressure, p, can be derived as follows:

$$\frac{p_a}{\rho_{a0}} = \frac{p_a + u_a}{\rho_a} = \frac{p_a + u_a + \Delta u_a}{\rho_a + \Delta \rho_a} = \text{constant} \Rightarrow \frac{p_a + u_a}{\rho_a} = \frac{\Delta u_a}{\Delta \rho_a} \tag{4.12}$$

And the relationship between $\Delta \rho_a$ and Δu_a can be obtained:

$$\Delta \rho_a = \frac{\rho_a}{p_a + u_a} \Delta u_a \tag{4.13}$$

Substituting Eq. (4.13) into Eq. (4.10) yields

$$\begin{aligned}\Delta u_a &= \frac{(p_a+u_a)\Delta q_a}{\rho_a n_a} - \frac{(p_a+u_a)\Delta n_a}{n_a} \\ &= \frac{p_a+u_a}{n_a}\left(\frac{\Delta q_a}{\rho_a} - \Delta n_a\right) \\ &= \frac{p_a+u_a}{n_a}\left(\frac{\Delta q_a}{\rho_a \Delta n_a} - 1\right)\Delta n_a \end{aligned} \tag{4.14}$$

Therefore, according to Eqs. (4.8) and (4.14), the incremental form of the pore-air pressure, Δu_a, can be rewritten as follows:

$$\Delta u_a = -(1-\xi)\frac{p_a+u_a}{n_a}\Delta n_a = -P\Delta n_a \tag{4.15}$$

where

$$P = (1-\xi)\frac{p_a+u_a}{n_a} \tag{4.16}$$

Rearranging Eq. (4.15) gives

$$\frac{\mathrm{d}(p_a+u_a)}{p_a+u_a} = -(1-\xi)\frac{\mathrm{d}n_a}{n_a} \tag{4.17}$$

And integrating Eq. (4.17) yields

$$\int \frac{1}{p_a+u_a}\mathrm{d}(p_a+u_a) = -(1-\xi)\int \frac{1}{n_a}\mathrm{d}n_a \Rightarrow \ln(\frac{p_a+u_a}{p_a}) = \ln\left(\frac{n_{a0}}{n_a}\right)^{(1-\xi)} \tag{4.18}$$

Then, according to Eq. (4.18), the relationship between the pore-air pressure, u_a, and the pore-air content, n_a, can be obtained as follows:

$$u_a = \left[\left(\frac{n_{a0}}{n_a}\right)^{1-\xi} - 1\right]p_a \tag{4.19}$$

For the above equation, when the soil is completely undrained, $\xi = 0$, and the equation will be degraded into Eq. (4.7); when the soil is completely drained, $\xi = 1$, and $u_a = 0$.

It should be noted that the sign convention in soil mechanics is the opposite of that in structural mechanics (or solid mechanics). However, the identical expression of Eq. (4.19) can be obtained no matter which of the two sign conventions is used. In order to correspond with the general mechanics analysis, the sign convention of structural mechanics will be adopted in the following formula derivations of this chapter (i.e., Sections 4.3 and 4.4) unless otherwise specified. And in accordance with this sign convention, the matric suction can be denoted as $s = u_w - u_a$. If the sign convention in soil mechanics is adopted, the matric suction should be express as $s = u_a - u_w$.

Then, according to Eq. (4.1), Δn_a can be obtained as follows:

$$\begin{aligned}\Delta n_a &= \frac{\partial n_a}{\partial S_r}\Delta S_r + \frac{\partial n_a}{\partial n}\Delta n \\ &= \frac{\partial n_a}{\partial S_r}\frac{\partial S_r}{\partial s}(\Delta u_w - \Delta u_a) + \frac{\partial n_a}{\partial n}\Delta n\end{aligned} \tag{4.20}$$

where

$$\Delta S_r = \frac{\partial S_r}{\partial s}\Delta s = \frac{\partial S_r}{\partial s}(\Delta u_w - \Delta u_a) \tag{4.21}$$

Substituting Eq. (4.15) into Eq. (4.20) and rearranging the equation yield the following expression for pore-air pressure, Δu_a:

$$\begin{aligned}\Delta u_a &= -\frac{P\frac{\partial n_a}{\partial S_r}\frac{\partial S_r}{\partial s}}{1 - P\frac{\partial n_a}{\partial S_r}\frac{\partial S_r}{\partial s}}\Delta u_w - \frac{P\frac{\partial n_a}{\partial n}}{1 - P\frac{\partial n_a}{\partial S_r}\frac{\partial S_r}{\partial s}}\Delta n \\ &= -\frac{P}{1 - P\frac{\partial n_a}{\partial S_r}\frac{\partial S_r}{\partial s}}\left(\frac{\partial n_a}{\partial S_r}\frac{\partial S_r}{\partial s}\Delta u_w + \frac{\partial n_a}{\partial n}\Delta n\right)\end{aligned} \tag{4.22}$$

To sum up, if the drainage condition of the pore air in the soil is known, the pore-air pressure, u_a, in the soil can be calculated by the pore-water pressure, u_w, and the soil porosity, n, according to Eq. (4.22). It is no longer necessary to solve the continuity equation of the pore-air phase. This simplification method makes the difficulty of solving the coupling equations of unsaturated seepage-consolidation greatly reduced.

4.3 The Effective Stress in the Simplified Consolidation Theory for Unsaturated Soils

In the simplified consolidation theory for unsaturated soils, the effective stress equation suggested by Bishop (1959) is adopted. The form of the equation is as follows:

$$\sigma' = (\sigma - u_a) + \chi(u_a - u_w) \tag{4.23}$$

where σ is the total normal stress; σ' is the effective normal stress; u_w is the pore-water pressure; u_a is the pore-air pressure; χ is the effective stress parameter, which represents the contribution of matric suction to the effective stress and can be obtained experimentally. The effective stress equation can be expressed as Eq. (4.23) either according to the sign convention of soil mechanics or according to that of structural mechannics.

According to the Bishop's effective stress equation, the total normal stress of soil can be expressed as

$$\sigma = \sigma' + u_a - \chi(u_a - u_w) \tag{4.24}$$

The incremental form of the total normal stress can be obtained as follows:

$$\Delta\sigma = \Delta\sigma' + \Delta u_a - \Delta\chi \cdot (-s) - \chi \cdot (\Delta u_a - \Delta u_w) \tag{4.25}$$

And $\Delta\chi$ can be calculated by

$$\Delta\chi = \frac{\partial\chi}{\partial s} \cdot \Delta s = \frac{\partial\chi}{\partial s}(\Delta u_w - \Delta u_a) \tag{4.26}$$

Substituting Eq. (4.26) into Eq. (4.25) yields

$$\Delta\sigma = \Delta\sigma' + \left(1 - \chi - \frac{\partial\chi}{\partial s} \cdot s\right) \cdot \Delta u_a + \left(\chi + \frac{\partial\chi}{\partial s} \cdot s\right) \cdot \Delta u_w \tag{4.27}$$

Considering that $\Delta n = \Delta\varepsilon_v$, the following equation can be obtained by substituting the expression of Δu_a [i.e., Eq. (4.22)] into Eq. (4.27):

$$\Delta\sigma = \Delta\sigma' + \left(\frac{\chi + \frac{\partial\chi}{\partial s}s - P\frac{\partial n_a}{\partial S_r}\frac{\partial S_r}{\partial s}}{1 - P\frac{\partial n_a}{\partial S_r}\frac{\partial S_r}{\partial s}}\right)\Delta u_w - \frac{(1 - \chi - \frac{\partial\chi}{\partial s}s)P\frac{\partial n_a}{\partial n}}{1 - P\frac{\partial n_a}{\partial S_r}\frac{\partial S_r}{\partial s}}\Delta\varepsilon_v \tag{4.28}$$

The above equation can be further written as:

$$\Delta\sigma = \Delta\sigma' + A_1\Delta u_w + A_2\Delta\varepsilon_v \tag{4.29}$$

where

$$A_1 = \frac{\chi + \frac{\partial\chi}{\partial s}s - P\frac{\partial n_a}{\partial S_r}\frac{\partial S_r}{\partial s}}{1 - P\frac{\partial n_a}{\partial S_r}\frac{\partial S_r}{\partial s}}; \quad A_2 = -\frac{(1 - \chi - \frac{\partial\chi}{\partial s}s)P\frac{\partial n_a}{\partial n}}{1 - P\frac{\partial n_a}{\partial S_r}\frac{\partial S_r}{\partial s}}.$$

The expression of Eq. (4.29) can also be derived using the sign convention in soil mechanics. However, A_1 and A_2 need to be expressed as follows:

$$A_1 = \frac{\chi + \frac{\partial\chi}{\partial s}s + P\frac{\partial n_a}{\partial S_r}\frac{\partial S_r}{\partial s}}{1 + P\frac{\partial n_a}{\partial S_r}\frac{\partial S_r}{\partial s}}\Delta u_w; A_2 = \frac{(1 - \chi - \frac{\partial\chi}{\partial s}s)P\frac{\partial n_a}{\partial n}}{1 + P\frac{\partial n_a}{\partial S_r}\frac{\partial S_r}{\partial s}}.$$

4.4 The Continuity Equation of Pore Water

4.4.1 The Flow of Water in Unsaturated Soils

The unsaturated soil has commonly been referred to as a three-phase system, including the soil particle phase, the water phase, and the air phase. The water phase and the air phase can be classified as fluids, and they will flow in the soil skeleton under the action of driving potential. A knowledge of the driving potentials is necessary for studying the seepage and consolidation in saturated and unsaturated soils. It is generally accepted that the hydraulic gradient is a basic driving potential that controls the flow of pore water. And the total water head, also called the total head or the hydraulic head, can effectively describe the characteristics of water flow in unsaturated soils (Lu and Likos 2004).

As with saturated soils, the total head in unsaturated soils can also be expressed as (according to the sign convention in soil mechanics)

$$h_w = z + \frac{u_w}{\rho_w g} \tag{4.30}$$

where z is the elevation head, and u_w is the pore-water pressure.

The total head in both saturated soils and unsaturated soils can be calculated by Eq. (4.30). In general conditions, the pore-water pressure in unsaturated soils is less than zero. But for some special conditions, u_a will appear greater than zero. For example, when an unsaturated soil with a relatively high degree of saturation is compressed, the excess pore-water pressure will be generated which may make u_w increase from a negative value to a positive value. Also, the pore water pressure in unsaturated soils is closely related to the soil type and the degree of saturation.

The flow of water in saturated soils is commonly described using Darcy's law. For a specific saturated soil, the hydraulic conductivity is relatively constant. Darcy's law also applies to the flow of water through unsaturated soils. However, the hydraulic conductivity in an unsaturated soil is no longer a constant, which varies with the degree of saturation (or the matric suction) of the soil. The Darcy's law for unsaturated soils is as follows:

$$v_x = -k_x \frac{\partial h_w}{\partial x} \tag{4.31}$$

where v_x is the flow rate of water in the x-direction; and k_x is the hydraulic conductivity of an unsaturated soil in the x-direction.

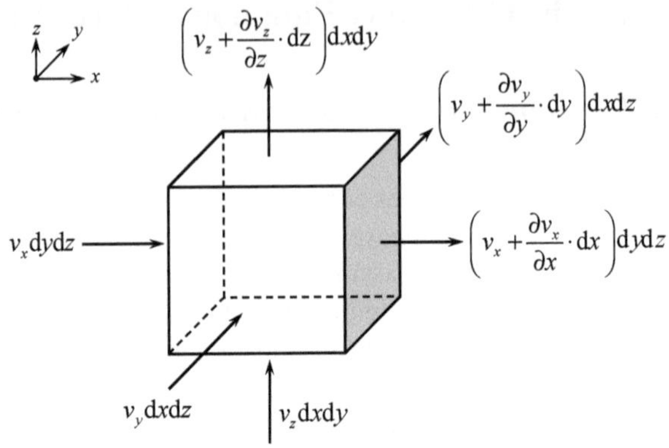

Fig. 4.1 Schematic of three-dimensional water flows

4.4.2 *Governing Equation for Saturated-Unsaturated Seepage*

The continuity equation of pore water in unsaturated soils can be determined by the conservation of water mass. That is, in a unit time, the amount of water that flows into an infinitesimal cubical element of unsaturated soil, dxdydz, is equal to the mass change of the element. Figure 4.1 shows the three-dimensional water flows in a soil element.

The amount of water that flows into the cubical element in a unit time can be expressed as

$$\Delta m_1 = -\rho\left(\frac{\partial v_x}{\partial x} + \frac{\partial v_y}{\partial y} + \frac{\partial v_z}{\partial z}\right)\mathrm{d}V \tag{4.32}$$

Meanwhile, for an unsaturated soil with a saturation degree of S_r, the mass change of water can be written as

$$\Delta m_2 = \frac{\partial(\rho\theta)}{\partial t}\mathrm{d}V = \frac{\partial(\rho n S_r)}{\partial t}\mathrm{d}V \tag{4.33}$$

According to the conservation of pore water mass in the soil, i.e., $\Delta m_1 = \Delta m_2$, the continuity equation of pore water can be expressed as follows:

$$\left(\frac{\partial v_x}{\partial x} + \frac{\partial v_y}{\partial y} + \frac{\partial v_z}{\partial z}\right) + \frac{1}{\rho}\cdot\frac{\partial(\rho n S_r)}{\partial t} = 0 \tag{4.34}$$

Equation (4.34) gives the general form of the continuity equation of pore water for both seepage analyses and consolidation analyses in unsaturated soils. When a soil is saturated, the degree of saturation of the soil, S_r, equals to 1, and Eq. (4.34) degrades into the continuity equation of pore water for saturated soils, as shown below:

$$\left(\frac{\partial v_x}{\partial x}+\frac{\partial v_y}{\partial y}+\frac{\partial v_z}{\partial z}\right)+\frac{1}{\rho}\cdot\frac{\partial(\rho n)}{\partial t}=0 \tag{4.35}$$

Due to the very small compressibility of water, the variation in water density, ρ, is usually neglected. The second term in the pore-water continuity equation [i.e., Eq. (4.34)] can be simplified and written as:

$$\frac{1}{\rho}\frac{\partial(\rho n S_r)}{\partial t}=S_r\frac{\partial n}{\partial t}+n\frac{\partial S_r}{\partial t} \tag{4.36}$$

Because of that $\frac{\partial n}{\partial t}=\frac{\partial \varepsilon_v}{\partial t}$, and $\frac{\partial S_r}{\partial t}=\frac{\partial S_r}{\partial s}\frac{\partial s}{\partial t}$, Eq. (4.36) can be further written as

$$\frac{1}{\rho}\frac{\partial(\rho n S_r)}{\partial t}=S_r\frac{\partial \varepsilon_v}{\partial t}+n\frac{\partial S_r}{\partial s}\frac{\partial s}{\partial t} \tag{4.37}$$

The continuity equation of pore water in seepage analyses and that in the coupling analysis of seepage-consolidation are usually treated differently. For seepage analyses, the volume change of soil skeleton is generally not considered, and the pore-air pressure keeps constant and equals to the atmospheric pressure. Therefore, the continuity equation of pore water in seepage analyses can be simplified:

$$\left(\frac{\partial v_x}{\partial x}+\frac{\partial v_y}{\partial y}+\frac{\partial v_z}{\partial z}\right)+n\frac{\partial S_r}{\partial \bar{h}}\frac{\partial \bar{h}}{\partial t}=0 \tag{4.38}$$

where $\bar{h}=\gamma_w h$.

For the coupling analyses of seepage-consolidation in unsaturated soils, both the volume change of soil skeleton and the pore-air pressure should be taken into account.

According to simplification method of the simplified consolidation theory, the pore-air pressure can be calculated by the following equation:

$$\Delta u_a=-\frac{P\frac{\partial n_a}{\partial S_r}\frac{\partial S_r}{\partial s}}{1-P\frac{\partial n_a}{\partial S_r}\frac{\partial S_r}{\partial s}}\Delta u_w-\frac{P\frac{\partial n_a}{\partial n}}{1-P\frac{\partial n_a}{\partial S_r}\frac{\partial S_r}{\partial s}}\Delta n \tag{4.39}$$

Then,

$$\frac{\partial u_a}{\partial t}=-\frac{P\frac{\partial n_a}{\partial S_r}\frac{\partial S_r}{\partial s}}{1-P\frac{\partial n_a}{\partial S_r}\frac{\partial S_r}{\partial s}}\frac{\partial u_w}{\partial t}-\frac{P\frac{\partial n_a}{\partial n}}{1-P\frac{\partial n_a}{\partial S_r}\frac{\partial S_r}{\partial s}}\frac{\partial n}{\partial t} \tag{4.40}$$

Substituting Eq. (4.40) into Eq. (4.37) yields

$$\begin{aligned}\frac{1}{\rho}\frac{\partial(\rho n S_r)}{\partial t}&=S_r\frac{\partial \varepsilon_v}{\partial t}+n\frac{\partial S_r}{\partial s}\left(-\frac{\partial \bar{h}}{\partial t}-\frac{\partial u_a}{\partial t}\right)\\&=\left(S_r+\frac{nP\frac{\partial n_a}{\partial n}\frac{\partial S_r}{\partial s}}{1-P\frac{\partial n_a}{\partial S_r}\frac{\partial S_r}{\partial s}}\right)\frac{\partial \varepsilon_v}{\partial t}+\left(-\frac{n\frac{\partial S_r}{\partial s}}{1-P\frac{\partial n_a}{\partial S_r}\frac{\partial S_r}{\partial s}}\right)\frac{\partial \bar{h}}{\partial t}\end{aligned} \tag{4.41}$$

Finally, the continuity equation of pore water in unsaturated seepage-consolidation analysis is derived as follows:

$$\left(\frac{\partial v_x}{\partial x}+\frac{\partial v_y}{\partial y}+\frac{\partial v_z}{\partial z}\right)+\tilde{S}_r\frac{\partial \varepsilon_v}{\partial t}+S_s\frac{\partial \bar{h}}{\partial t}=0 \tag{4.42}$$

where

$$\tilde{S}_r=S_r+\frac{nP\frac{\partial n_a}{\partial n}\frac{\partial S_r}{\partial s}}{1-P\frac{\partial n_a}{\partial S_r}\frac{\partial S_r}{\partial s}};\quad S_s=n\beta S_r-\frac{n\frac{\partial S_r}{\partial s}}{1-P\frac{\partial n_a}{\partial S_r}\frac{\partial S_r}{\partial s}}.$$

By substituting Darcy's law into Eq. (4.42), the continuity equation of pore water can be further expressed as follows:

$$-\frac{1}{\gamma_w}\left[\frac{\partial}{\partial x}\left(k_x\frac{\partial \bar{h}}{\partial x}\right)+\frac{\partial}{\partial y}\left(k_y\frac{\partial \bar{h}}{\partial y}\right)+\frac{\partial}{\partial z}\left(k_z\frac{\partial \bar{h}}{\partial z}\right)\right]+\tilde{S}_r\frac{\partial \varepsilon_v}{\partial t}+S_s\frac{\partial \bar{h}}{\partial t}=0 \tag{4.43}$$

The expression of Eq. (4.43) can also be derived using the sign convention in soil mechanics. However, $\tilde{S}_r$ and S_s need to be expressed as

$$\tilde{S}_r=-S_r+\frac{nP\frac{\partial n_a}{\partial n}\frac{\partial S_r}{\partial s}}{1+P\frac{\partial n_a}{\partial S_r}\frac{\partial S_r}{\partial s}};\quad S_s=n\beta S_r-\frac{n\frac{\partial S_r}{\partial s}}{1+P\frac{\partial n_a}{\partial S_r}\frac{\partial S_r}{\partial s}}.$$

4.5 Coupling Analysis Method for Unsaturated Consolidation

The core wall is a key part of the earth-rockfill dam. The stress-deformation and seepage behaviors of the core material in high earth-rockfill dams are significantly different from those of low dams. In high dams, the core material bears extremely complex changes in seepage and stress-deformation state in the process of construction, impounding, and operation.

Due to the complexity of the construction conditions and the limitations of the test conditions and numerical methods, the existing numerical simulation analysis methods cannot fully reflect the actual states of stress, deformation, and pore-water pressure of the dam materials. In addition, there are complex interactions between stress-deformation field, the physical-state field and the seepage field in the dam body. In particular, for the problems of saturated-unsaturated seepage and consolidation analysis in high earth-rock dams, the effect of multi-physics coupling should not be neglected. The current theories and methods still hardly reflect these characteristics. For the consolidation problems with high stress and large deformations, ignoring the coupling effects will cause significant errors in the numerical analysis.

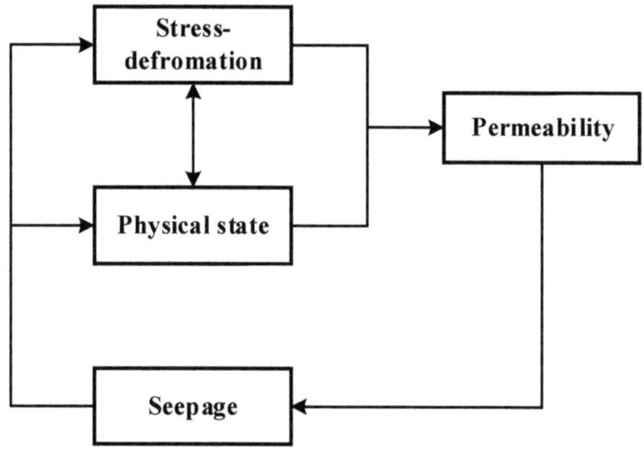

Fig. 4.2 Diagram of the multi-physics coupling relationship in the core wall of high earth-rock dams

The complex interactions between the stress field, the deformation field, the physical-state (including the density, the degree of saturation, the mesostructure, etc.) field, and seepage field in the core wall of high earth-rockfill dams can be illustrated by Fig. 4.2. The permeability of soil directly affects the distribution of the seepage field and plays a decisive role in the evolution of the pore-water pressure in the core wall. And the variations in the seepage field will change the stress and deformation state and physical state of the soil. There are also interactions between the stress-strain state and the physical state of soil; the physical state (e.g., the degree of saturation) will affect the matrix suction in the soil and thus affects the effective stress; soil deformations (e.g., compression) will induce variations in the physical state of soil. And both the stress-deformation state and the physical state have significant influences on the permeability of soil.

Therefore, in the multi-physics coupling analysis, the following issues should be studied in detail: (1) a reasonable consolidation theory for saturated and unsaturated soils; (2) an effective simulation method for hydraulic conductivity; (3) the calculation of effective stress in saturated-unsaturated soils; (4) the stress-strain relationship of soil skeleton.

In order to simulate the interactions in the core wall more reasonably, a multi-physics coupling consolidation analysis method for unsaturated soils is developed based on the traditional consolidation theory. The governing equations are shown in Eq. (4.44), which contains the equilibrium equations of soil skeleton, the continuity equation of pore water, and the equation of hydraulic conductivity.

$$\left.\begin{array}{l}\left\{\begin{array}{l}\frac{\partial \sigma_x}{\partial x}+\frac{\partial \tau_{xy}}{\partial y}+\frac{\partial \tau_{xz}}{\partial z}=0\\ \frac{\partial \tau_{yx}}{\partial x}+\frac{\partial \sigma_y}{\partial y}+\frac{\partial \tau_{yz}}{\partial z}=0\\ \frac{\partial \tau_{zx}}{\partial x}+\frac{\partial \tau_{zy}}{\partial y}+\frac{\partial \sigma_z}{\partial z}+F_g=0\end{array}\right.\\ -\frac{1}{\gamma_w}\left[\frac{\partial}{\partial x}\left(k_x\frac{\partial \bar{h}}{\partial x}\right)+\frac{\partial}{\partial y}\left(k_y\frac{\partial \bar{h}}{\partial y}\right)+\frac{\partial}{\partial z}\left(k_z\frac{\partial \bar{h}}{\partial z}\right)\right]+\tilde{S}_r\frac{\partial \varepsilon_V}{\partial t}+S_s\frac{\partial \bar{h}}{\partial t}=0\\ k=f(\varepsilon_v,\sigma,S_r)\end{array}\right\} \tag{4.44}$$

In the above model of coupling analysis, in addition to the equilibrium equations and the continuity equation of pore water, the following important relations are also included.

(1) Stress-strain relationship in unsaturated soils

One of the important properties of soil is the nonlinear relationship between the stress and strain. In the consolidation analysis for earth-rock dams, the stress-strain relationship of soil is often described by the equation of an incremental form, as shown in Eq. (4.45). For this incremental model, different numerical solutions (e.g., incremental method and iteration method) can be adopted in the solving process of consolidation calculation.

$$\left\{\mathrm{d}\sigma'\right\}=[D]_t\{\mathrm{d}\varepsilon\} \tag{4.45}$$

where $[D]_t$ is the incremental form of the stiffness matrix.

In seepage and consolidation analysis, the soil is usually assumed to be a homogeneous and isotropic material. For an isotropic material, there will be only two independent variables in the matrix, $[D]_t$. And the stress-strain relationship can be express as follows:

$$\left\{\begin{array}{c}\Delta\sigma_x\\ \Delta\sigma_y\\ \Delta\sigma_z\\ \Delta\sigma_{xy}\\ \Delta\sigma_{yz}\\ \Delta\sigma_{zx}\end{array}\right\}=\left[\begin{array}{cccccc}d_1 & d_2 & d_2 & 0 & 0 & 0\\ & d_1 & d_2 & 0 & 0 & 0\\ & & d_1 & 0 & 0 & 0\\ & & & d_3 & 0 & 0\\ & sym. & & & d_3 & 0\\ & & & & & d_3\end{array}\right]_t\left\{\begin{array}{c}\Delta\varepsilon_x\\ \Delta\varepsilon_y\\ \Delta\varepsilon_z\\ \Delta\gamma_{xy}\\ \Delta\gamma_{yz}\\ \Delta\gamma_{zx}\end{array}\right\} \tag{4.46}$$

where d_3 can be further calculated by d_1 and d_2, as follows:

$$d_3=\frac{1}{2}(d_1-d_2) \tag{4.47}$$

For a transverse isotropy material, there are five independent variables in the matrix, $[D]_t$. And the stress-strain relationship can be expressed as follows:

$$\begin{Bmatrix} \Delta\sigma_x \\ \Delta\sigma_y \\ \Delta\sigma_z \\ \Delta\sigma_{xy} \\ \Delta\sigma_{yz} \\ \Delta\sigma_{zx} \end{Bmatrix} = \begin{bmatrix} d_{11} & d_{12} & d_{13} & 0 & 0 & 0 \\ & d_{11} & d_{13} & 0 & 0 & 0 \\ & & d_{33} & 0 & 0 & 0 \\ & & & d_{44} & 0 & 0 \\ & sym. & & & d_{44} & 0 \\ & & & & & d_{66} \end{bmatrix}_t \begin{Bmatrix} \Delta\varepsilon_x \\ \Delta\varepsilon_y \\ \Delta\varepsilon_z \\ \Delta\gamma_{xy} \\ \Delta\gamma_{yz} \\ \Delta\gamma_{zx} \end{Bmatrix} \tag{4.48}$$

where $d_{66} = \frac{1}{2}(d_{11} - d_{12})$.

For a general anisotropic material, there are 21 independent variables in the matrix $[D]_t$, and the stress-strain relationship can be expressed as follows:

$$\begin{Bmatrix} \Delta\sigma_x \\ \Delta\sigma_y \\ \Delta\sigma_z \\ \Delta\sigma_{xy} \\ \Delta\sigma_{yz} \\ \Delta\sigma_{zx} \end{Bmatrix} = \begin{bmatrix} d_{11} & d_{12} & d_{13} & d_{14} & d_{15} & d_{16} \\ & d_{22} & d_{23} & d_{24} & d_{25} & d_{26} \\ & & d_{33} & d_{34} & d_{35} & d_{36} \\ & & & d_{44} & d_{45} & d_{46} \\ & sym. & & & d_{55} & d_{56} \\ & & & & & d_{66} \end{bmatrix}_t \begin{Bmatrix} \Delta\varepsilon_x \\ \Delta\varepsilon_y \\ \Delta\varepsilon_z \\ \Delta\gamma_{xy} \\ \Delta\gamma_{yz} \\ \Delta\gamma_{zx} \end{Bmatrix} \tag{4.49}$$

(2) The effective stress in unsaturated soils

In this dissertation, the effective stress concept proposed by Bishop is adopted to describe the stress state in unsaturated soils. The effective stress controls the mechanical behavior of a soil (i.e., the deformation and strength behavior). For unsaturated soils, the effective stress consists of certain combinations of the net normal stress and the matric suction. Figure 4.3a gives the normal stresses in a soil element which include the net normal stresses and the matric suctions. Figure 4.3b shows the effective stresses in the soil element.

(3) Hydraulic conductivity model for pore water

The hydraulic conductivity of a clayey soil is greatly affected by the physical and mechanical states of the soil. Meanwhile, the hydraulic conductivity has decisive effects on the consolidation process, such as the generation and dissipation of the excess pore-water pressure, the stress and deformation, and so on. Therefore, in the unsaturated consolidation analysis of high earth-rock dams, the influence of stress, deformation, and physical state on the hydraulic conductivity should be fully considered, as shown in the following equation:

$$k = f(\varepsilon_v, \sigma, S_r) \tag{4.50}$$

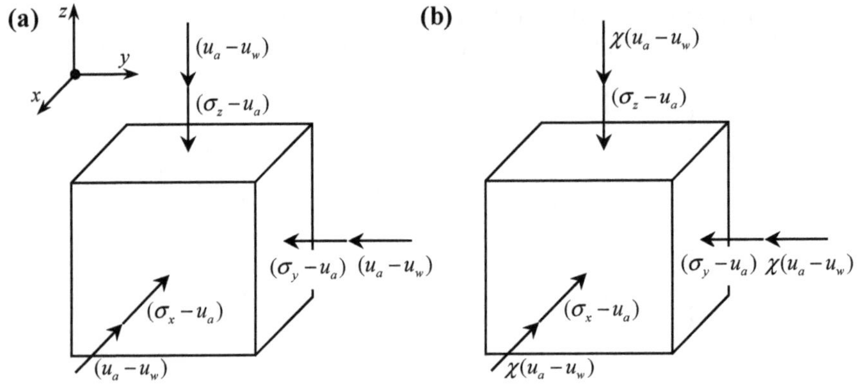

Fig. 4.3 Normal stresses on a cubical soil element of infinitesimal dimensions: **a** normal stresses in unsaturated soil; **b** effective stresses in unsaturated soil

4.6 Description of Constitutive Relations and Parameters

4.6.1 Soil Skeleton Stress-Strain Relationship

At present, the constitutive models commonly used to describe the stress-strain relationship of soils include multiple categories, such as the nonlinear elastic models, the elasto-plastic models, the constitutive models for special soils, etc. Among them, the Duncan and Chang's EB model (Duncan and Chang 1970), as one of the nonlinear elastic models, is adopted in this dissertation. This model is simple, and its parameters have clear physical meanings and are easy to determine experimentally. It has been widely used in geotechnical engineering, especially in the stress-deformation analyses of earth-rock dams.

4.6.1.1 Duncan and Chang's EB Model

In this study, the Duncan and Chang's EB model is used to describe the stress-strain relationship of the saturated dam materials. In the model, the tangent modulus, E_t, and the bulk modulus, B, can be expressed as follows:

$$E_t = Kp_a\left(\frac{\sigma_3}{p_a}\right)^n (1 - R_f S_l)^2 \tag{4.51}$$

$$B = K_b p_a\left(\frac{\sigma_3}{p_a}\right)^m \tag{4.52}$$

where S_l is the shear stress level of the soil and can be calculated by Eq. (4.53).

$$S_l = \frac{(1 - \sin\varphi)(\sigma_1 - \sigma_3)}{2c \cdot \cos\varphi + 2\sigma_3 \cdot \sin\varphi} \tag{4.53}$$

Then, the elastic matrix, $[D]$, in the constitutive relations of soil skeleton can be expressed as

$$[D] = \frac{3B}{9B - E_t}\begin{bmatrix} 3B + E_t & 3B - E_t & 3B - E_t & 0 & 0 & 0 \\ 3B - E_t & 3B + E_t & 3B - E_t & 0 & 0 & 0 \\ 3B - E_t & 3B - E_t & 3B + E_t & 0 & 0 & 0 \\ 0 & 0 & 0 & E_t & 0 & 0 \\ 0 & 0 & 0 & 0 & E_t & 0 \\ 0 & 0 & 0 & 0 & 0 & E_t \end{bmatrix} \tag{4.54}$$

In Duncan and Chang's EB model, in order to describe the plastic deformations of the soil during loading and unloading, the unloading-reloading modulus is calculated as follows:

$$E_{ur} = K_{ur}p_a\left(\frac{\sigma_3}{p_a}\right)^n \tag{4.55}$$

It can be seen that there are eight parameters in this model, that is, c, φ, K, K_{ur}, n, R_{f}, K_b, and m, which can be evaluated using a group of conventional triaxial tests.

The Mohr-Coulomb envelopes for almost all soils are curved to some extent. In general, the wider the range of involved confining pressure, the greater the curvature will be, especially for cohesionless soils such as sand, gravel, and rockfill. In this study, for the rockfill materials, the cohesion, c, takes the value of 0, and the nonlinearity of the strength envelope is given by:

$$\varphi = \varphi_0 - \Delta\varphi \log\left(\frac{\sigma_3}{p_a}\right) \tag{4.56}$$

where φ_0 and $\Delta\varphi$ are the nonlinear strength parameters.

The Duncan and Chang's EB model employs the new unloading-reloading modulus to describe the plastic deformations of soil. However, this model still does not exceed the category of the elastic theory. In practical engineering applications, the soil is usually under complex stress-path. Thus, estimating the loading-unloading state becomes one of the important issues in using the model. In view of this, the following stress state function was defined:

$$SS = S_l \cdot \left(\frac{\sigma_3}{p_a}\right)^{1/4} \tag{4.57}$$

The maximum value of SS in the entire stress history of a soil is recorded as SS_m, and the stress level, S_c, normalized by the current stress, σ_3, can be expressed as:

$$S_c = \frac{SS_m}{(\sigma_3/p_a)^{1/4}} \tag{4.58}$$

Then the current stress level, S_l, is compared with S_c to determine the loading-unloading-reloading state as well as the value of the tangent modulus, E_t'. Detailed procedures are as follows:

When $S_l \geq S_c$, the stress state is loading, and the tangent modulus will be

$$E_t' = E_t \tag{4.59}$$

When $S_l \leq 0.75S_c$, the stress state is unloading or reloading, and the tangent modulus will be

$$E_t' = E_{ur} \tag{4.60}$$

When $S_c > S_l > 0.75S_c$, the tangent modulus will be calculated by interpolation as follows:

$$E_t' = E_t + \frac{S_c - S_l}{0.25S_c}(E_{ur} - E_t) \tag{4.61}$$

4.6.1.2 Constitutive Relations for Unsaturated Soils

Seeking for the constitutive models describing the stress-strain relationship of unsaturated soils, the simplest way is to extend the constitutive models of saturated soils to unsaturated soils. The deformation and strength properties of soils are controlled by the effective stress in the soils. Based on this premise, it can be inferred that the stress-strain relationship of a soil is also controlled by the effective stress and is independent of whether the soil is in a saturated or unsaturated state. In this way, when the effective stress in unsaturated soils is determined, the constitutive models of saturated soils can be extended to use in unsaturated soils.

There are different ways for the extension of the constitutive models. In this chapter, according the suggestion of Shen (2000), the Duncan-Chang's EB model is extended to unsaturated soils based on the Bishop's principle of effective stress. The stress in the model is replaced by the effective stress defined by the Bishop's equation, and the tangent modulus, E_t, and the bulk modulus, B, in the model can be expressed as follows:

$$E_t = Kp_a\left(\frac{\sigma_3^* + \chi s}{p_a}\right)^n (1 - R_f S_l)^2 \tag{4.62}$$

$$B = K_b p_a \left(\frac{\sigma_3^* + \chi s}{p_a} \right)^m \tag{4.63}$$

In this way, the constitutive models of saturated soils can be easily applied to the consolidation analysis of unsaturated soils. The feasibility of this method has been approved by many scholars.

4.6.2 *Hydraulic Conductivity Models*

The flow of water in saturated soils is commonly described using Darcy's law. In this study, it is assumed that Darcy's law also applies to the flow of water through unsaturated soils. In order to illustrate the interactions between the multi-physics, the variation of the soil permeability with the stress, strain, and physical-state should be considered. In this section, the models of hydraulic conductivity for saturated and unsaturated soils used in this study are briefly described.

4.6.2.1 Hydraulic Conductivity of Saturated Soils

The variations of stress-strain and physical state will induce significant changes in the saturated hydraulic conductivity of the clayey soil. Lei et al. (2016) studied the seepage characteristics of the clayey soil under shear conditions based on a series of triaxial seepage tests. It was found that the change of void ratio induced by volumetric compression and the mesostructural change induced by shear deformation are the main factors that lead to the variation of the saturated hydraulic conductivity in the process of shear deformation. On the basis of the test results, a mathematical model was then developed to describe the permeability of soil:

$$k_s = c \cdot \exp(ae + bS_l) \tag{4.64}$$

where k_s is the saturated hydraulic conductivity of the soil (the unit of k_s is cm/s); e is the void ratio; S_l is the shear stress level in the range of 0–1.0; a, b, and c are the undetermined constants of the model (c has the same unit as k_s).

Equation (4.64) can be further rewritten as:

$$k_s = c \exp(ae) \cdot \exp(bS_l) \tag{4.65}$$

where the first part, $c \exp(ae)$, reflects the influence of the volumetric strain on the hydraulic conductivity of the soil; the second part, $\exp(bS_l)$, reflects the influence of the structural changes induced by the shear action on the hydraulic conductivity of the soil.

According to the mathematical model, Fig. 4.4 shows the general form of the spatial surface in three-dimensional coordinates formed by the void ratio, e, the

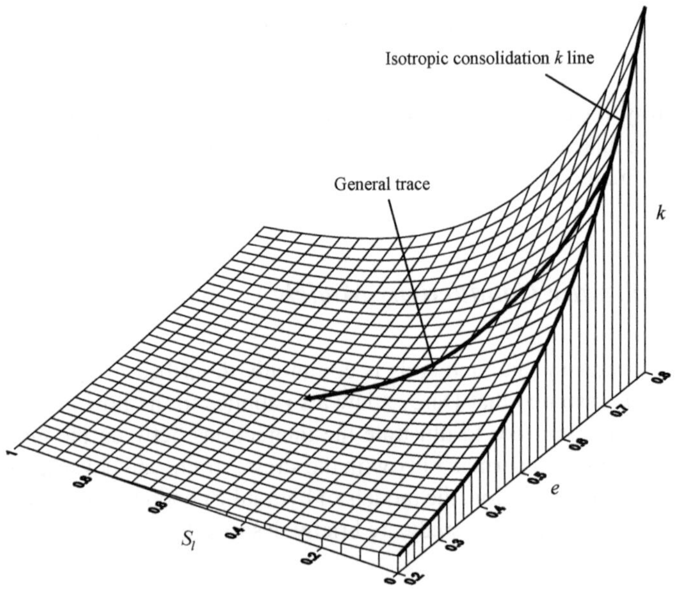

Fig. 4.4 Space form of the mathematical model of hydraulic conductivity

shear stress level, S_l, and the hydraulic conductivity, k_s, of the soil. Every point on this spatial surface corresponds to a hydraulic conductivity of the soil with a specific void ratio and shear stress level. As shown in the figure, during the shear process of a soil, the variation of the hydraulic conductivity can be described by a general curve in the spatial surface, which is named "general trace". When the shear stress level equals zero ($S_l = 0$), the soil only experiences the isotropic consolidation stage without shear deformation, and Eq. (4.66) can be degraded into

$$k_s = c\exp(ae) \tag{4.66}$$

In this isotropic consolidation condition, the hydraulic conductivity of the soil will vary along the "isotropic consolidation k line".

4.6.2.2 Hydraulic Conductivity of Unsaturated Soils

For unsaturated soils, the hydraulic conductivity is closely related to the variables of the void ratio, the degree of saturation, etc. In general, the effect of the degree of saturation on soil permeability is far greater than that of other facts such as the void ratio and the stress state. The permeability of a soil will change drastically from a saturated state to an unsaturated state. In the pore system of a soil, water can only flow through the pore space filled with water. As the soil becomes unsaturated, air replaces part of the water in the pores, and this causes the water to flow through the

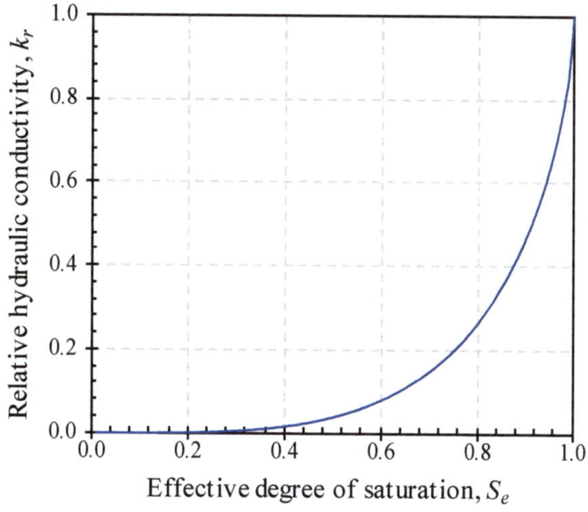

Fig. 4.5 Relationship between the relative hydraulic conductivity, k_r, and the effective degree of saturation, S_e

smaller pores with an increased tortuosity to the flow path. A further decrease in water content leads to a further decrease in the flow paths. As a result, the hydraulic conductivity of pore water will significantly decrease. For most of the soils, the hydraulic conductivity will change more than six orders of magnitude when the soil changes from the saturated state to a lower degree of saturation (Lu and Likos 2004). Figure 4.5 shows the relationship between the relative hydraulic conductivity, k_r, and the effective degree of saturation, S_e.

In this study, an unsaturated hydraulic conductivity model proposed by Mualem (1976) is adopted. The equations of the model are as follows:

$$\left.\begin{aligned} k &= k_s \cdot k_r \\ k_r &= S_e^{\frac{1}{2}}\left[1-\left(1-S_e^{\frac{1}{m}}\right)^m\right]^2 \end{aligned}\right\} \tag{4.67}$$

where k_s is the saturated hydraulic conductivity; k_r is the relative hydraulic conductivity in the range of 0–1.0; S_e is the effective degree of saturation; m is a parameter from the van Genuchten model of the soil-water characteristic curve.

The effective degree of saturation, S_e, can be expressed in terms of the volumetric water content, θ, the saturated volumetric water content, θ_s, and the residual volumetric water content, θ_r, as follows:

$$S_e = \frac{\theta - \theta_r}{\theta_s - \theta_r} \tag{4.68}$$

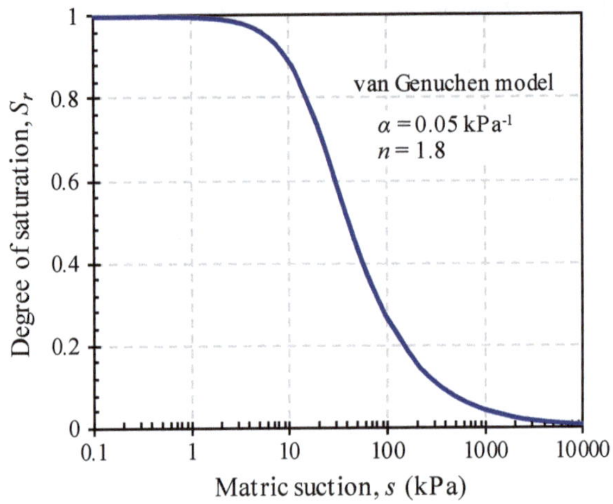

Fig. 4.6 Soil-water characteristic curve modeled using the van Genuchten model

4.6.3 Soil-Water Characteristic Curve

The soil-water characteristic curve (SWCC) is broadly defined as the relationship between the water content and the matric suction of a soil, which characterizes the basic properties of the soil. Numerous approaches have been proposed to model the soil-water characteristic curve. The van Genuchten (1980) model is one of the famous SWCC models, which has been commonly used for geotechnical engineering applications. And it is adopted in the coupling consolidation analysis of this dissertation. The van Genuchten model is shown as follows:

$$S_e = \left[1 + (\alpha \cdot s)^n\right]^{-m} \tag{4.69}$$

where s is the matric suction; S_e is the effective degree of saturation; α, n, and m are the fitting parameters of the model. The α parameter is related to the air-entry condition, where α approximates the inverse of the air-entry pressure. The n parameter is related to the pore size distribution of the soil, which describes the desorption rate of pore water. The m parameter is usually constrained as $m = 1 - 1/n$, or $m = 1 - 1/(2n)$.

Figure 4.6 shows the soil-water characteristic curve modeled using the van Genuchten model. The model parameters for the curve are as follows: $\alpha = 0.5\ \text{kPa}^{-1}$ and $n = 1.8$. A soil with a relatively high air-entry pressure is characterized by a smaller value of α parameter. And a soil with a greater desorption rate (i.e., with a steeper characteristic curve) is characterized by a larger value of n parameter. To illustrate the difference, the model parameters of sandy soils are compared with the parameters of clayey soils. The air-entry pressure of a clayey soil is higher than that of a sandy soil, then the a parameter for the clayey soil is greater than that for the sandy soil. And the desorption rate of a clayey soil is smaller than that of a sandy soil, then the n parameter of the clayey soil will be smaller than that of a sandy soil.

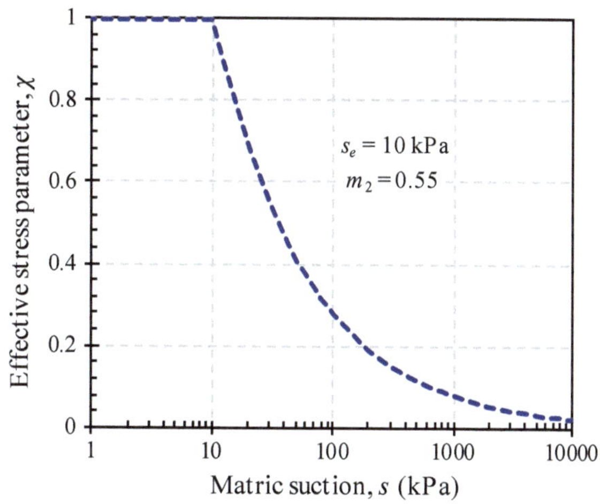

Fig. 4.7 Relationship between the effective stress parameter, χ, and the matric suction, s, modeled using the Shen model

4.6.4 Effective Stress Parameter in Unsaturated Soils

In the unsaturated soil mechanics, the effective stress parameter (also known as the reduction coefficient) is one of the most important unsaturated parameters, and it has a decisive impact on the results of deformation analysis and strength analysis of unsaturated soils. The physical meaning of the "reduced suction" can be understood as the portion of the matric suction that is effective for the deformation and strength of a soil (Chen et al. 2008). Shen (2003) proposed that the relationship between the effective stress parameter, χ, and the matric suction, s, can be expressed by the following equation:

$$\left.\begin{array}{ll} \chi = \left(\frac{s}{s_e}\right)^{-m_2} & s \geq s_e \\ \chi = 1 & s < s_e \end{array}\right\} \tag{4.70}$$

where s_e is the air-entry pressure, which is also denoted as $(u_a - u_w)_b$; m_2 is the fitting parameter of the equation.

Based on the model in Eq. (4.70), Figs. 4.7 and 4.8 plot the curve of the effective stress parameter, χ, versus the matric suction, s, and the curve of the reduced suction, χs, versus the matric suction, s, respectively. The values of the model parameters for the curves are: s_e = 10 kPa, m_2 = 0.55. According to the effective stress parameter model, the soil with a higher suction stress, s, will have a smaller value of effective stress parameter, χ, and a greater value of reduced suction, χs.

The effective stress parameter, χ, can be determined based on either the principle of deformation equivalence or the principle of strength equivalence. The following briefly describes the method to obtain the effective stress parameter using these two

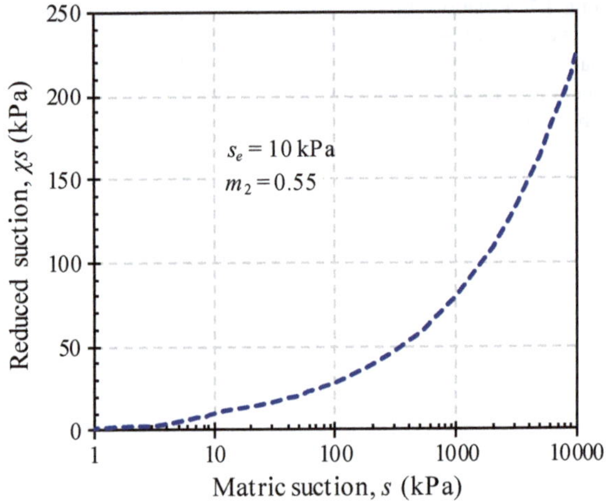

Fig. 4.8 Relationship between the reduced suction, $\chi(u_a - u_w)$, and the matric suction, s, modeled using the Shen model

equivalence principles. And based on the triaxial test data in Chap. 3, the effective stress parameters of the gravelly clay material in Nuozhadu dam is obtained.

4.6.4.1 Deformation Equivalence Method

The method of determining the effective stress parameters using the principle of equivalence deformation is shown in Fig. 4.9 (Shen 2003). In this method, two identical samples of saturated soils are prepared to conduct a consolidation test and a drying shrinkage test, respectively. In the consolidation test, the soil shrinks under the effect of effective stress, that is, the void ratio, e, decreases with the increase of the consolidation stress, p, as shown by Curve #1 in Fig. 4.9. Then, the drying shrinkage test is carried out in the same initial condition as the consolidation test. The void ratio will change with the change of the matric suction as shown by Curve #2 in Fig. 4.9. When the soil samples of the two tests reach the same void ratio, e_0 (as shown by points A and B in the figure), the deformations as well as the corresponding stress states of the soil in the two tests should be the same. According to the Bishop's principle of effective stress, the following equation holds:

$$(\sigma_A - u_{aA}) + \chi(\sigma_{aA} - u_{wA}) = (\sigma_B - u_{aB}) + \chi(\sigma_{aB} - u_{wB}) \tag{4.71}$$

The left side of the equation represents the effective stress of the consolidation test, in which $\sigma_A = p_A$ and $u_{aA} = u_{wA} = 0$. The right side of the equation is the effective stress of the drying shrinkage test, where $\sigma_B = 0$, $u_{aB} = 0$, and $s_B = u_{aB} - u_{wB}$. Then the equation can be further expressed as

$$p_A = \chi s_B \tag{4.72}$$

Fig. 4.9 Determination of the reduced suction and effective stress parameter

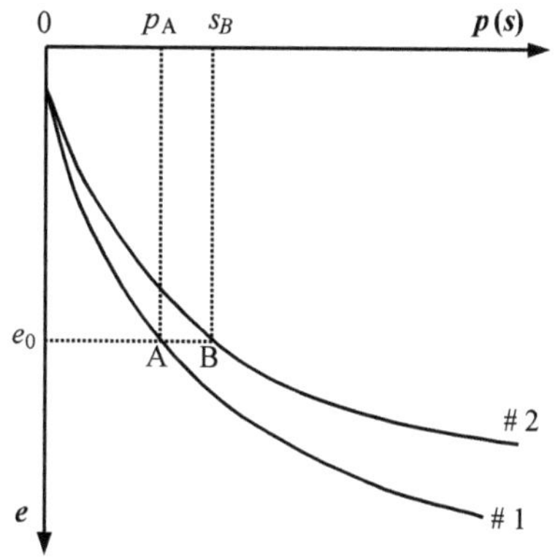

Therefore, when the matric suction is s_B (point B in the figure), the corresponding effective stress parameter can be calculated by

$$\chi = \frac{p_A}{s_B} \tag{4.73}$$

Through the above tests, the value of the effective stress parameter at any matric suction can be determined.

4.6.4.2 Strength Equivalence Method

The strength equivalence method is similar to the deformation equivalence method (Khalili and Khabbaz 1998). In this method, the contribution of the reduced suction to the strength of an unsaturated soil is estimated by comparing the strength of the unsaturated soil and that of the saturated soil. It is assumed that the soil strength parameters (e.g., c' and φ') will not change with the change of the matric suction. Relevant test results have shown that when the matric suction changes, there is no significant change in φ' (Drumright 1991; Escario and Sáez 1986).

According to the Bishop's principle of effective stress, the drained shear strength of an unsaturated soil can be written as

$$\tau = c' + [(\sigma - u_a) + \chi(u_a - u_w)] \tan \varphi' \tag{4.74}$$

where u_a is the pore-air pressure, and $u_a = 0$.

On the other hand, the drained shear strength of a saturated soil can be expressed as

$$\tau_s = c' + (\sigma - u_w) \tan \varphi' \tag{4.75}$$

Considering that in the consolidated drained test, there is $u_a = u_w = 0$. Then Eq. (4.75) can be further written as

$$\tau_s = c' + (\sigma - u_a)\tan\varphi' \tag{4.76}$$

When the normal stress, σ, in Eq. (4.74) is equal to that in Eq. (4.76), the difference between the two equations will be the contribution of matric suction to the strength of the unsaturated soil, as follows:

$$\tau - \tau_s = \chi(u_a - u_w)\tan\varphi' \tag{4.77}$$

Therefore, when the matric suction equals to $(u_a - u_w)$, the effective stress parameter can be calculated by the following equation:

$$\chi = \frac{\tau - \tau_s}{(u_a - u_w)\tan\varphi'} \tag{4.78}$$

4.6.4.3 Analysis of the Effective Stress Parameter for the Nuozhadu Core Material

In Chap. 3 of this dissertation, the gravelly clay material of the Nuozhedu high earth-rockfill dam is used to conduct the saturated and unsaturated triaxial CD tests. Based on the test data from Chap. 3, the effective stress parameter is analyzed using the strength equivalence method, and then the value of the undetermined parameter in the effective stress parameter model is obtained. In the following analysis, the influence of stress path (i.e., changing of the matric suction during shear) on the strength is ignored.

The air-entry pressure of the gravelly clay of the Nuozhadu dam takes 40 kPa. And the drained strength parameters from the triaxial CD tests are as follows: c = 63.8 kPa and φ = 27.9°.

The data of the two sets of triaxial CD tests under different confining pressures are analyzed to obtain the strength difference, $(\tau - \tau_s)$, between the saturated soil and the unsaturated soil, as shown in Table 4.1. It can be seen that with the increase of the confining pressure, the strength difference between the unsaturated soil and the saturated soil shows a decreasing trend.

The test points under different confining pressures are marked on the coordinate system of χ *versus*$(u_a - u_w)/(u_a - u_w)_b$, as shown in Fig, 4.10. It can be seen that the relationship is generally linear in log-log scale. Then, the mathematical expression of the effective stress parameter is obtained by linear fitting:

$$\chi = \left(\frac{s}{s_e}\right)^{-0.37} \tag{4.79}$$

Table 4.1 Test results of $(\tau - \tau_s)$ under different confining pressures

Confining pressure, σ_3 (kPa)	$\tau - \tau_s$ (kPa)
100	93.63
200	81.18
400	71.30
800	54.18
1300	56.01
2000	38.34

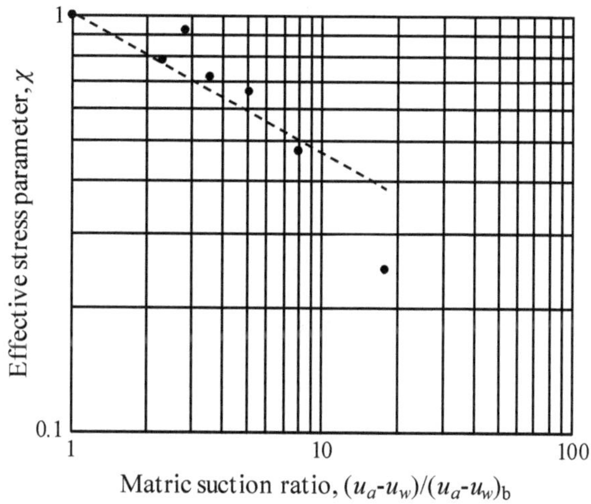

Fig. 4.10 Relationship between the effective stress parameter, χ, and the matric suction ratio, $(u_a - u_w)/(u_a - u_w)_b$, from the test results

4.7 Summary

The characteristics of stress-deformation and permeability of the core wall in high earth-rock dams are significantly different from that of low dams and show an obvious multi-physics coupling effect. During construction, impounding and operation of the dams, the core material bears complex changes in the stress-strain state, physical state, and the seepage state. In this chapter, the multi-physics coupling issues of unsaturated soils are studied, and the following conclusions are obtained:

(1) The governing equations of the simplified consolidation theory for unsaturated soils are deduced, and a multi-physics coupling analysis method is proposed on the basis of the simplified consolidation theory. In this coupling analysis method, the interactions between the physical state field, the stress-deformation field, and the seepage field are taken into account, and a more reasonable mode for saturated-unsaturated seepage-consolidation analysis is established.

(2) The core issues in the multi-physics coupling analysis method for saturated-unsaturated soils, such as the calculation of soil effective stress, the simulation of soil permeability, the stress-strain relationship, and so on, are analyzed, and the solutions to these issues are given.

(3) Based on the results of the saturated-unsaturated triaxial tests, the unsaturated effective stress parameter of the gravelly clay of the Nuozhadu dam is obtained using the strength equivalence method. This parameter will then be adopted in the unsaturated consolidation analysis of the Nuozhadu high earth-rockfill dam.

References

Bishop AW (1959) The principle of effective stress. Teknisk Ukeblad 39:859–863

Chen TL, Chen SS, Zhang WM, Wang NX (2008) Application of reduced suction to earth pressure calculation of unsaturated soils and swell increment. Chin J Rock Mechan Eng S2:3341–3348

Coleman JD (1962) Stress/strain relations for partly saturated soils. Geotechnique 12(4):348–350

Deng G (2005) Systematic research on distortion mechanism of over consolidated clay slope. Doctoral dissertation, Tsinghua University, Beijing

Drumright EE (1991) The contribution of matric suction to the shear strength of unsaturated soils. Doctoral dissertation, Colorado State University, Fort Collins

Duncan JM, Chang CY (1970) Nonlinear analysis of stress and strain in soils. J Soil Mech Foundations Div 96(5):1629–1653

Escario V, Sáez J (1986) The shear strength of partly saturated soils. Géotechnique 36(3):453–456

Jie YX, Wen YF, Deng G, Chen R, Xu ZP (2012) Impact of soil deformation on phreatic line in earth-fill dams. Comput Geosci 46:44–50

Khalili N, Khabbaz MH (1998) A unique relationship for χ for the determination of the shear strength of unsaturated soils. Geotechnique 48(5):681–687

Lei HJ, Wu YK, Yu YZ, Zhang BY, Lv H (2016) Influence of shear on permeability of clayey soil. Int J Geomech 16(5):04016010

Lu N, Likos WJ (2004) Unsaturated soil mechanics. Wiley, New Jersey

Mualem Y (1976) A new model for predicting the hydraulic conductivity of unsaturated porous media. Water Resour Res 12(3):513–522

Shen ZJ (2000) Theoretical soil mechanics. China Water Power Press, Beijing

Shen ZJ (2003) Simplified consolidation theory for unsaturated soils and its application. Hydroscience Eng 4:1–6

Shen ZJ (2006) Exploitation of practical use of unsaturated soil mechanics. Chin J Geotech Eng 28(2):256–259

van Genuchten MT (1980) A closed-form equation for predicting the hydraulic conductivity of unsaturated soils. Soil Sci Soc Am J 44(5):892–898

Chapter 5
Numerical Realization of the Unsaturated Seepage-Consolidation Theory

The coupling analysis method of seepage and consolidation for saturated-unsaturated soils is proposed in Chap. 4. On the basis of this analysis method, the numerical realization of unsaturated seepage and consolidation for complex engineering problems is studied in this chapter. The finite element method (FEM) equations of saturated-unsaturated consolidation problems are derived using the Galerkin method, and a finite element program is then developed for numerical computation. The development process of the finite element program and the common boundary conditions in seepage and consolidation analysis are explained in detail. Finally, the FE program is applied to analyze typical examples of earth-rock dams and compared with the traditional saturated analysis program to verify the coupling analysis method as well as the program.

5.1 FEM Equations for the Consolidation Theory of Unsaturated Soils

Biot's consolidation theory (Biot 1955, 1941), as a representative of the consolidation theories for saturated soils, has been widely applied in geotechnical engineerings. For unsaturated soils, the consolidation theory is much more complex, and more factors need to be considered in unsaturated consolidation analyses. In the multi-physics coupling analysis method for unsaturated soils proposed in Chap. 4, the governing equations include the equilibrium equations, the continuity equation, the effective stress equation, the stress-strain relation, and the strain-displacement relation, which are similar to the saturated consolidation theories. In this method, the difficulty of solving the unsaturated consolidation problems will be greatly reduced, which makes it possible to be applied in coupling consolidation analyses of large-scale engineerings.

Y. Wu, *Multi-physics Coupling Analysis of Clayey Core Wall of High Earth-Rockfill Dam*, Springer Theses, https://doi.org/10.1007/978-981-13-1032-4_5

The partial differential equations of the seepage-consolidation theory cannot be directly solved by the analytical method (Wang 2003). Numerical methods are the only viable solution to the equations. In this chapter, the finite element method is adopted, and the FEM equations for programming is derived using the Galerkin method.

In practice, almost all the physical problems in geotechnical engineering can be described by differential equations. Then the engineering problems are transformed into a mathematical problem. The Galerkin method, as one of the weighted residual methods (WRM), has been widely used in the finite element method (Wang 2003; Zienkiewicz and Taylor 1977; Chandrupatla et al. 2002; Feng et al. 2014). The weighted residual method is an approximate method for solving the differential equations based on the equations as well as the boundary conditions. The idea of the weighted residual method is to convert the differential equations into their corresponding equivalent integral form and find an approximate solution based on the integral form to make that the weighted integral of the residuals in the domain is zero. In this method, the weight functions are chosen to be identical to the trial functions.

For a system of partial differential equations, if an unknown function $\boldsymbol{u}$ is assumed to be a solution to the equation system in the domain Ω, the following condition should be satisfied:

$$\boldsymbol{A}(\boldsymbol{u}) = \begin{Bmatrix} A_1(\boldsymbol{u}) \\ A_2(\boldsymbol{u}) \\ \vdots \end{Bmatrix} = \mathbf{0}, \quad \boldsymbol{u} \in \Omega \tag{5.1}$$

On the boundary Γ of the domain Ω, the following boundary condition is also satisfied:

$$\boldsymbol{B}(\boldsymbol{u}) = \begin{Bmatrix} B_1(\boldsymbol{u}) \\ B_2(\boldsymbol{u}) \\ \vdots \end{Bmatrix} = \mathbf{0}, \quad \boldsymbol{u} \in \Gamma \tag{5.2}$$

It is assumed that the approximate solution of the unknown function $\boldsymbol{u}$ in the domain Ω is as follows:

$$\boldsymbol{u} \approx \hat{\boldsymbol{u}} = \sum_{i=1}^{n} \boldsymbol{N}_i \boldsymbol{a}_i = \boldsymbol{N}\boldsymbol{a} \tag{5.3}$$

where $\boldsymbol{N}_i$ is the trial function (also known as the shape function), and $\boldsymbol{a}_i$ is the undetermined parameter.

For the system of differential equations with a given boundary condition, its equivalent integral form expressed by the Galerkin method is as follows:

$$\int_{\Omega} \boldsymbol{N}_j^T \boldsymbol{A}(\boldsymbol{u}) \mathrm{d}\Omega - \int_{\Gamma} \boldsymbol{N}_j^T \boldsymbol{B}(\boldsymbol{u}) \mathrm{d}\Gamma = \boldsymbol{0} \quad (j = 1, 2, \ldots, n) \tag{5.4}$$

Equation (5.4) can also be written in an approximate form:

$$\int_{\Omega} \boldsymbol{N}_j^T \boldsymbol{A}(\boldsymbol{N}\boldsymbol{a}) \mathrm{d}\Omega - \int_{\Gamma} \boldsymbol{N}_j^T \boldsymbol{B}(\boldsymbol{N}\boldsymbol{a}) \mathrm{d}\Gamma = \boldsymbol{0} \quad (j = 1, 2, \ldots, n) \tag{5.5}$$

The above gives the basic mathematical principle of the Galerkin method. Then combined with the partial integration method, Eqs. (5.4) and (5.5) can be further processed to reduce the order of differential and obtain the corresponding natural boundary conditions.

Based on this principle, the governing equations for saturated-unsaturated seepage and consolidation analysis can be discretized in space. The derivation process of the FEM equations is given below.

5.1.1 *Finite Element Discretization for the Continuity Equation*

The continuity equation of pore water for seepage and consolidation analyses of saturated-unsaturated soils is as follows:

$$-\frac{1}{\gamma_w}\left[\frac{\partial}{\partial x}\left(k_x \frac{\partial \bar{h}}{\partial x}\right) + \frac{\partial}{\partial y}\left(k_y \frac{\partial \bar{h}}{\partial y}\right) + \frac{\partial}{\partial z}\left(k_z \frac{\partial \bar{h}}{\partial z}\right)\right] + \tilde{S}_r \frac{\partial \varepsilon_{\mathrm{v}}}{\partial t} + S_s \frac{\partial \bar{h}}{\partial t} = 0 \tag{5.6}$$

where $\varepsilon_{\mathrm{v}} = \varepsilon_x + \varepsilon_y + \varepsilon_z$.

In the coupling analysis method proposed in this study, the coupling effect between the stress-deformation, the physical state, and the permeability of the soil is taken into account. In the continuity equation of the unsaturated seepage-consolidation theory, the hydraulic conductivity, k, is not constant but varies with the change of the soil state. The soil permeability is described by a mathematical model of hydraulic conductivity, which characterizes the complex relationship between the hydraulic conductivity, k, and the shear stress level, S_l, the void ratio, e, and the degree of saturation, S_r. To solve the continuity equation of the coupling equations, a nonlinear iterative algorithm will be adopted. The hydraulic conductivity, k, is continuously updated and corrected in the iteration process until a relatively accurate value is approached. Thus, for each nonlinear iteration step, the hydraulic conductivity of the soil will be regarded as a constant. Then, Eq. (5.6) can be expressed as follows:

$$-\frac{1}{\gamma_w}\left(k_x \frac{\partial^2 \bar{h}}{\partial x^2} + k_y \frac{\partial^2 \bar{h}}{\partial y^2} + k_z \frac{\partial^2 \bar{h}}{\partial z^2}\right) + \tilde{S}_r \frac{\partial \varepsilon_{\mathrm{v}}}{\partial t} + S_s \frac{\partial \bar{h}}{\partial t} = 0 \tag{5.7}$$

Moreover, the flux boundary condition of the pore-water continuity equation can be expressed as

$$lv_x + mv_y + nv_z = v_n \tag{5.8}$$

where v_x, v_y, v_z are the flow velocity in the directions of x, y, and z, respectively, which can be calculated by Darcy's law; v_n is the flow velocity in the normal direction of the boundary; $\{l\ m\ n\}^T$ is the normal direction of the boundary surface.

The domain is then discretized into small elements with simple shapes. The shape function of the element is denoted by N_i. Then the pore-water pressure at any position inside the element can be expressed by the element interpolation functions as:

$$\bar{h} = \sum_{i=1}^{m} N_i \bar{h}_i \tag{5.9}$$

For any weight function N_i, the equivalent integral forms of Eqs. (5.7) and (5.8) yield

$$\int_V N_i \left[-\frac{1}{\gamma_w} \left(k_x \frac{\partial^2 \bar{h}}{\partial x^2} + k_y \frac{\partial^2 \bar{h}}{\partial y^2} + k_z \frac{\partial^2 \bar{h}}{\partial z^2} \right) + \tilde{S}_r \frac{\partial \varepsilon_v}{\partial t} + S_s \frac{\partial \bar{h}}{\partial t} \right] \mathrm{d}V = 0 \tag{5.10}$$

$$\int_A N_i \left(lv_x + mv_y + nv_z - v_n \right) \mathrm{d}A = 0 \tag{5.11}$$

The items in Eq. (5.10) can be further derived using the partial integration method as follows:

$$\begin{cases} \int_V \left(N_i k_x \frac{\partial^2 \bar{h}}{\partial x^2} \right) \mathrm{d}V = -\int_V \left(k_x \frac{\partial N_i}{\partial x} \frac{\partial \bar{h}}{\partial x} \right) \mathrm{d}V + \int_A \left(N_i k_x \frac{\partial \bar{h}}{\partial x} l \right) \mathrm{d}A \\ \int_V \left(N_i k_y \frac{\partial^2 \bar{h}}{\partial y^2} \right) \mathrm{d}V = -\int_V \left(k_y \frac{\partial N_i}{\partial y} \frac{\partial \bar{h}}{\partial y} \right) \mathrm{d}V + \int_A \left(N_i k_y \frac{\partial \bar{h}}{\partial y} m \right) \mathrm{d}A \\ \int_V \left(N_i k_z \frac{\partial^2 \bar{h}}{\partial z^2} \right) \mathrm{d}V = -\int_V \left(k_z \frac{\partial N_i}{\partial z} \frac{\partial \bar{h}}{\partial z} \right) \mathrm{d}V + \int_A \left(N_i k_z \frac{\partial \bar{h}}{\partial z} n \right) \mathrm{d}A \end{cases} \tag{5.12}$$

Substituting Eq. (5.12) into Eq. (5.10) yields

$$\begin{aligned} &\frac{1}{\gamma_w} \int_V \left(k_x \frac{\partial N_i}{\partial x} \frac{\partial \bar{h}}{\partial x} + k_y \frac{\partial N_i}{\partial y} \frac{\partial \bar{h}}{\partial y} + k_z \frac{\partial N_i}{\partial z} \frac{\partial \bar{h}}{\partial z} \right) \mathrm{d}V + \int_V \tilde{S}_r N_i \frac{\partial \varepsilon_v}{\partial t} dV \\ &+ \int_V S_s N_i \frac{\partial \bar{h}}{\partial t} \mathrm{d}V = \frac{1}{\gamma_w} \int_A N_i \left(k_x \frac{\partial \bar{h}}{\partial x} l + k_y \frac{\partial \bar{h}}{\partial y} m + k_z \frac{\partial \bar{h}}{\partial z} n \right) \mathrm{d}A \end{aligned} \tag{5.13}$$

Moreover, according to Darcy's law, the equivalent integral equation of the element boundary shown in Eq. (5.11) can be further written as follows:

$$-\int_A N_i \frac{1}{\gamma_w}\left(k_x \frac{\partial \bar{h}}{\partial x} l + k_y \frac{\partial \bar{h}}{\partial y} m + k_z \frac{\partial \bar{h}}{\partial z} n\right) dA = \int_A N_i v_n dA \tag{5.14}$$

Combining Eqs. (5.13) and (5.14) gives

$$\frac{1}{\gamma_w}\int_V \left(k_x \frac{\partial N_i}{\partial x}\frac{\partial \bar{h}}{\partial x} + k_y \frac{\partial N_i}{\partial y}\frac{\partial \bar{h}}{\partial y} + k_z \frac{\partial N_i}{\partial z}\frac{\partial \bar{h}}{\partial z}\right) dV$$
$$+ \int_V \tilde{S}_r N_i \frac{\partial \varepsilon_v}{\partial t} dV + \int_V S_s N_i \frac{\partial \bar{h}}{\partial t} dV = -\int_A N_i v_n dA \tag{5.15}$$

This equation is the FE form of the pore-water continuity equation obtained by using the Galerkin method combined with the flux boundary condition. Equation (5.15) can be further written as matrixes:

$$\int_V [B_s]^T [k][B_s] dV \{\bar{h}\} + \int_V \tilde{S}_r [\bar{N}]^T \{M\}^T [B] dV \{\dot{\delta}\}$$
$$+ \int_V S_s [\bar{N}]^T [\bar{N}] dV \{\dot{\bar{h}}\} = -\int_A [\bar{N}]^T v_n \mathrm{d}A \tag{5.16}$$

where

$$[B_s] = \begin{bmatrix} \frac{\partial N_1}{\partial x} & \frac{\partial N_2}{\partial x} & \frac{\partial N_3}{\partial x} & & \frac{\partial N_m}{\partial x} \\ \frac{\partial N_1}{\partial y} & \frac{\partial N_2}{\partial y} & \frac{\partial N_3}{\partial y} & \cdots & \frac{\partial N_m}{\partial y} \\ \frac{\partial N_1}{\partial z} & \frac{\partial N_2}{\partial z} & \frac{\partial N_3}{\partial z} & & \frac{\partial N_m}{\partial z} \end{bmatrix};$$

$$\{M\} = \begin{bmatrix} 1 & 1 & 1 & 0 & 0 & 0 \end{bmatrix}^T;$$

$$\{\varepsilon\} = \begin{bmatrix} \varepsilon_x & \varepsilon_y & \varepsilon_z & \gamma_{xy} & \gamma_{yz} & \gamma_{xz} \end{bmatrix}^T;$$

$$[k] = \frac{1}{\gamma_w} \cdot \begin{bmatrix} k_x & 0 & 0 \\ 0 & k_y & 0 \\ 0 & 0 & k_z \end{bmatrix};$$

$$[\bar{N}] = [\, N_1 \; N_2 \; N_3 \; \ldots \; N_m \,].$$

Therefore, the finite element formulation of the continuity equation is as follows:

$$[K_c]^{T'}\{\dot{\delta}\} + [K_s]\{\bar{h}\} + [K_p]\{\dot{\bar{h}}\} = -\int_A [\bar{N}]^T v_n \mathrm{d}A \tag{5.17}$$

where

$$[K_s] = \int_V [B_s]^T[k][B_s]dV;$$

$$[K_c]^{T'} = \int_V \tilde{S}_r[\bar{N}]^T\{M\}^T[B]dV;$$

$$[K_p] = \int_V S_s[\bar{N}]^T[\bar{N}]dV.$$

The equation can be further discretized over time, and its incremental form in a time increment of Δt can be express as

$$\begin{aligned}&[K_c]^{T'}\{\Delta\delta\} + \Delta t\left(\theta[K_s]_n\{\bar{h}_n\} + (1-\theta)[K_s]_{n-1}\{\bar{h}_{n-1}\}\right)\\&\quad + [K_p](\{\bar{h}_n\} - \{\bar{h}_{n-1}\}) = -\Delta t\int_A [\bar{N}]^T[\theta v_n + (1-\theta)v_{n-1}]dA\end{aligned} \tag{5.18}$$

where θ is an integral parameter. To ensure the stability of numerical calculations, the value of the parameter should satisfy $\theta \geq 0.5$ (Booker and Small 1975). And in general, the preferred value is $\theta = 2/3$.

Equation (5.18) can be further rearranged into the following form:

$$\begin{aligned}&[K_c]^{T'}\{\Delta\delta\} + (\theta\Delta t[K_s] + [K_p])\{\bar{h}\}\\&= -\Delta t\{R_q\} + [K_p]\{\bar{h}_{n-1}\} - \Delta t(1-\theta)[K_s]_{n-1}\{\bar{h}_{n-1}\}\end{aligned} \tag{5.19}$$

where $\{R_q\} = \int_A [\bar{N}]^T[\theta v_n + (1-\theta)v_{n-1}]dA$.

5.1.2 *Finite Element Discretization for the Equilibrium Equations*

5.1.2.1 Equilibrium Equations Regarding the Total Stress

To be consistent with conventional mechanical analysis, the sign convention of structural mechanics, which is opposite to that in soil mechanics, is adopted in this section. By equilibrium analysis for a soil element, the equilibrium equations expressed by total stress can be obtained as follows:

$$\begin{cases} \frac{\partial \sigma_x}{\partial x} + \frac{\partial \tau_{xy}}{\partial y} + \frac{\partial \tau_{xz}}{\partial z} = 0 \\ \frac{\partial \tau_{yx}}{\partial x} + \frac{\partial \sigma_y}{\partial y} + \frac{\partial \tau_{yz}}{\partial z} = 0 \\ \frac{\partial \tau_{zx}}{\partial x} + \frac{\partial \tau_{zy}}{\partial y} + \frac{\partial \sigma_z}{\partial z} + F_g = 0 \end{cases} \tag{5.20}$$

where F_g is the self-weight load of the soil (usually taken as the natural unit weight or the saturated unit weight).

And the stress boundary condition of the soil element is as follows:

$$\begin{cases} n_x \cdot \sigma_x + n_y \cdot \tau_{xy} + n_z \cdot \tau_{xz} = \bar{F}_x \\ n_x \cdot \tau_{yx} + n_y \cdot \sigma_y + n_z \cdot \tau_{yz} = \bar{F}_y \\ n_x \cdot \tau_{zx} + n_y \cdot \tau_{zy} + n_z \cdot \sigma_z = \bar{F}_z \end{cases} \tag{5.21}$$

where $\bar{F}$ is the force on the boundary surface.

The shape function of the soil element is denoted by N_i. Then the displacement at any position within the element can be expressed by the following interpolation function:

$$u = \sum_{i=1}^{m} N_i u_i \tag{5.22}$$

Taking the shape function N_i as the weight function, the equivalent integral form of the equilibrium equations within the soil element will be as follows:

$$\begin{cases} \int_V N_i \left(\frac{\partial \sigma_x}{\partial x} + \frac{\partial \tau_{xy}}{\partial y} + \frac{\partial \tau_{xz}}{\partial z} \right) \mathrm{d}V = 0 \\ \int_V N_i \left(\frac{\partial \tau_{yx}}{\partial x} + \frac{\partial \sigma_y}{\partial y} + \frac{\partial \tau_{yz}}{\partial z} \right) \mathrm{d}V = 0 \\ \int_V N_i \left(\frac{\partial \tau_{zx}}{\partial x} + \frac{\partial \tau_{zy}}{\partial y} + \frac{\partial \sigma_z}{\partial z} + F_g \right) \mathrm{d}V = 0 \end{cases} \tag{5.23}$$

And the equivalent integral form of the boundary condition will be

$$\begin{cases} \int_A N_i \left(l \cdot \sigma_x + m \cdot \tau_{xy} + n \cdot \tau_{xz} - \bar{F}_x \right) \mathrm{d}A = 0 \\ \int_A N_i \left(l \cdot \tau_{yx} + m \cdot \sigma_y + n \cdot \tau_{yz} - \bar{F}_y \right) \mathrm{d}A = 0 \\ \int_A N_i \left(l \cdot \tau_{zx} + m \cdot \tau_{zy} + n \cdot \sigma_z - \bar{F}_z \right) \mathrm{d}A = 0 \end{cases} \tag{5.24}$$

The equilibrium equations shown in Eq. (5.23) can be further derived using the partial integration method (divergence theorem):

$$\begin{cases} \int\limits_V N_i \frac{\partial \sigma_x}{\partial x} \mathrm{d}V = \iiint N_i \frac{\partial \sigma_x}{\partial x} \mathrm{d}x\mathrm{d}y\mathrm{d}z = \iint N_i \sigma_x \mathrm{d}y\mathrm{d}z - \int\limits_V \sigma_x \frac{\partial N_i}{\partial x} \mathrm{d}V \\ \int\limits_V N_i \frac{\partial \tau_{xy}}{\partial x} \mathrm{d}V = \iint N_i \tau_{xy} \mathrm{d}y\mathrm{d}z - \int\limits_V \tau_{xy} \frac{\partial N_i}{\partial x} \mathrm{d}V \\ \int\limits_V N_i \frac{\partial \tau_{xz}}{\partial x} \mathrm{d}V = \iint N_i \tau_{xz} \mathrm{d}y\mathrm{d}z - \int\limits_V \tau_{xz} \frac{\partial N_i}{\partial x} \mathrm{d}V \end{cases} \tag{5.25}$$

Substituting Eq. (5.25) into the first equation in Eq. (5.23) yields

$$\begin{aligned} &\iint \left(N_i \sigma_x \mathrm{d}y\mathrm{d}z + N_i \tau_{xy} \mathrm{d}x\mathrm{d}z + N_i \tau_{xz} \mathrm{d}x\mathrm{d}y\right) \\ &\quad - \int\limits_V \left(\sigma_x \frac{\partial N_i}{\partial x} + \tau_{xy} \frac{\partial N_i}{\partial y} + \tau_{xz} \frac{\partial N_i}{\partial z}\right) \mathrm{d}V = 0 \end{aligned} \tag{5.26}$$

Then Eq. (5.26) is further rearranged as follows:

$$\int\limits_V \left(\sigma_x \frac{\partial N_i}{\partial x} + \tau_{xy} \frac{\partial N_i}{\partial y} + \tau_{xz} \frac{\partial N_i}{\partial z}\right) \mathrm{d}V = \iint \left(N_i \sigma_x \mathrm{d}y\mathrm{d}z + N_i \tau_{xy} \mathrm{d}x\mathrm{d}z + N_i \tau_{xz} \mathrm{d}x\mathrm{d}y\right) \tag{5.27}$$

By the same token, the weak forms of the equivalent integral for the equilibrium equations in the y and z directions are obtained:

$$\int\limits_V \left(\tau_{yx} \frac{\partial N_i}{\partial x} + \sigma_y \frac{\partial N_i}{\partial y} + \tau_{yz} \frac{\partial N_i}{\partial y}\right) \mathrm{d}V = \iint \left(N_i \tau_{yx} \mathrm{d}y\mathrm{d}z + N_i \sigma_y \mathrm{d}x\mathrm{d}z + N_i \tau_{yz} \mathrm{d}x\mathrm{d}y\right) \tag{5.28}$$

$$\begin{aligned} &\int\limits_V \left(\tau_{zx} \frac{\partial N_i}{\partial x} + \tau_{zy} \frac{\partial N_i}{\partial y} + \sigma_z \frac{\partial N_i}{\partial y} - N_i F_g\right) \mathrm{d}V \\ &= \iint \left(N_i \tau_{zx} \mathrm{d}y\mathrm{d}z + N_i \tau_{zy} \mathrm{d}x\mathrm{d}z + N_i \sigma_z \mathrm{d}x\mathrm{d}y\right) \end{aligned} \tag{5.29}$$

In addition, according to the integral form of the surface force boundary condition shown in Eq. (5.24) yields

$$\begin{cases} \iint N_i(\sigma_x \mathrm{d}y\mathrm{d}z + \tau_{xy}\mathrm{d}x\mathrm{d}z + \tau_{xz}\mathrm{d}x\mathrm{d}y) = \int\limits_A N_i \bar{F}_x \mathrm{d}A \\ \iint N_i(\tau_{yx} \mathrm{d}y\mathrm{d}z + \sigma_y\mathrm{d}x\mathrm{d}z + \tau_{yz}\mathrm{d}x\mathrm{d}y) = \int\limits_A N_i \bar{F}_y \mathrm{d}A \\ \iint N_i(\tau_{zx} \mathrm{d}y\mathrm{d}z + \tau_{zy}\mathrm{d}x\mathrm{d}z + \sigma_z\mathrm{d}x\mathrm{d}y) = \int\limits_A N_i \bar{F}_z \mathrm{d}A \end{cases} \tag{5.30}$$

On the basis of Eqs. (5.27)–(5.30), the following equations can be obtained:

$$\begin{cases} \int\limits_V \left(\sigma_x \frac{\partial N_i}{\partial x} + \tau_{xy}\frac{\partial N_i}{\partial y} + \tau_{xz}\frac{\partial N_i}{\partial z}\right)\mathrm{d}V = \int\limits_A N_i \bar{F}_x \mathrm{d}A \\ \int\limits_V \left(\tau_{yx} \frac{\partial N_i}{\partial x} + \sigma_y\frac{\partial N_i}{\partial y} + \tau_{yz}\frac{\partial N_i}{\partial y}\right)\mathrm{d}V = \int\limits_A N_i \bar{F}_y \mathrm{d}A \\ \int\limits_V \left(\tau_{zx} \frac{\partial N_i}{\partial x} + \tau_{zy}\frac{\partial N_i}{\partial y} + \sigma_z\frac{\partial N_i}{\partial y}\right)\mathrm{d}V = \int\limits_A N_i \bar{F}_z \mathrm{d}A + \int\limits_V N_i F_g \mathrm{d}V \end{cases} \tag{5.31}$$

Hence, the equilibrium equations in terms of total stress in a discrete soil element can be expressed as follows:

$$\int\limits_V [B]^T\{\sigma\}\mathrm{d}V = \int\limits_A [N]^T\{\bar{F}\}\mathrm{d}A + \int\limits_V [N]^T\{F\}\mathrm{d}V \tag{5.32}$$

where

$$[B] = [B_1, B_2, \ldots, B_m], [B_i] = \begin{bmatrix} \frac{\partial N_i}{\partial x} & 0 & 0 \\ 0 & \frac{\partial N_i}{\partial y} & 0 \\ 0 & 0 & \frac{\partial N_i}{\partial z} \\ \frac{\partial N_i}{\partial y} & \frac{\partial N_i}{\partial x} & 0 \\ 0 & \frac{\partial N_i}{\partial z} & \frac{\partial N_i}{\partial y} \\ \frac{\partial N_i}{\partial z} & 0 & \frac{\partial N_i}{\partial x} \end{bmatrix};$$

$$\{\sigma\} = \begin{Bmatrix} \sigma_x \\ \sigma_y \\ \sigma_z \\ \tau_{xy} \\ \tau_{yz} \\ \tau_{xz} \end{Bmatrix}; \ [N]=\begin{bmatrix} N_1 & & & & N_m & & \\ & N_1 & & \cdots & & N_m & \\ & & N_1 & & & & N_m \end{bmatrix};$$

$$\{\bar{F}\}=\begin{Bmatrix} \bar{F}_x \\ \bar{F}_y \\ \bar{F}_z \end{Bmatrix}; \quad \{F\}=\begin{Bmatrix} F_1 \\ F_2 \\ F_3 \end{Bmatrix}=\begin{Bmatrix} 0 \\ 0 \\ F_g \end{Bmatrix}.$$

5.1.2.2 Equilibrium Equations Regarding the Effective Stress

The finite element formulations of the equilibrium equations have been derived on the basis of the total-stress form. According to the principle of effective stress for saturated-unsaturated soils, the effective-stress form of the finite element formulations can be further obtained, as shown below.

According to Eq. (5.32), the finite element form of the equilibrium equations in terms of total stress can be expressed as

$$\int_V [B]^T \{\sigma\} \mathrm{d}V = \{R_F\} \tag{5.33}$$

where

$$\{R_F\} = \int_A [N]^T \{\bar{F}\} \mathrm{d}A + \int_V [N]^T \{F\} \mathrm{d}V; \quad \{\bar{F}\}=\begin{Bmatrix} \bar{F}_x \\ \bar{F}_y \\ \bar{F}_z \end{Bmatrix}; \quad \{F\}=\begin{Bmatrix} F_1 \\ F_2 \\ F_3 \end{Bmatrix}=\begin{Bmatrix} 0 \\ 0 \\ F_g \end{Bmatrix}.$$

For the problems of material nonlinearity, in order to simulate the stress-strain behavior of soils under multi-stage loading, the equilibrium equation in Eq. (5.33) can be expressed in an incremental form:

$$\int_V [B]^T \{\Delta\sigma\} \mathrm{d}V = \{\Delta R_F\} \tag{5.34}$$

where $\{\Delta R_F\} = \int_A [N]^T \{\Delta \bar{F}\} \mathrm{d}A + \int_V [N]^T \{\Delta F\} \mathrm{d}V$.

The simplified consolidation theory for unsaturated soils has been illustrated in Chap. 4. Based on the theory, the total stress can be converted into an incremental form:

$$\Delta\sigma = \Delta\sigma' + A_1\Delta u_w + A_2\Delta\varepsilon_v \tag{5.35}$$

Then, the above equation can be further written as tensors:

$$\begin{aligned}\{\Delta\sigma\} &= \left\{\Delta\sigma'\right\} + A_1\{M\}\Delta u_w + A_2\{M\}\{M\}^T\{\Delta\varepsilon\} \\ &= [D]\{\Delta\varepsilon\} + A_2\{M\}\{M\}^T\{\Delta\varepsilon\} + A_1\{M\}\Delta u_w \\ &= [\bar{D}]\{\Delta\varepsilon\} + A_1\{M\}\Delta u_w\end{aligned} \tag{5.36}$$

where

$$[\bar{D}] = [D] + A_2\{M\}\{M\}^T;$$
$$\{M\} = \left\{1\ 1\ 1\ 0\ 0\ 0\right\}^T.$$

Substituting Eq. (5.36) into Eq. (5.34) and considering that $\Delta\bar{h} = -\Delta u_w$ (derived from $\bar{h} = \gamma_w z - u_w$) yield:

$$\int\limits_V [B]^T[\bar{D}][B]\{\Delta\delta\}\mathrm{d}V - A_1\int\limits_V [B]^T\{M\}\Delta\bar{h}\mathrm{d}V=\{\Delta R_F\} \tag{5.37}$$

And further,

$$\int\limits_V [B]^T[\bar{D}][B]\{\Delta\delta\}\mathrm{d}V - A_1\int\limits_V [B]^T\{M\}[\bar{N}]\left\{\Delta\bar{h}\right\}\mathrm{d}V=\{\Delta R_F\} \tag{5.38}$$

Thus, the effective-stress form of the equilibrium equation can be expressed as follows:

$$[K]\{\Delta\delta\} - A_1[K_c]\left\{\Delta\bar{h}\right\} = \{\Delta R_F\} \tag{5.39}$$

where

$$[K] = \int\limits_V [B]^T[\bar{D}][B]\mathrm{d}V;$$

$$[K_c] = \int\limits_V [B]^T\{M\}[\bar{N}]\mathrm{d}V;$$

$$\{\Delta R_F\} = \int\limits_A [N]^{\mathrm{T}}\{\Delta\bar{F}\}\mathrm{d}A + \int\limits_V [N]^{\mathrm{T}}\{\Delta F\}\mathrm{d}V.$$

Considering $\Delta\bar{h} = \bar{h}_n - \bar{h}_{n-1} = \bar{h} - \bar{h}_{n-1}$, the above equation can be written as:

$$[K]\{\Delta\delta\} - A_1[K_c]\{\bar{h}\} = \{\Delta R_F\} - A_1[K_c]\{\bar{h}_{n-1}\} \tag{5.40}$$

To sum up, Eqs. (5.19) and (5.40) are the finite element formulations of the pore-water continuity equation and the equilibrium equations, respectively. Based on the combination of the two equations, the coupling consolidation problems of saturated-unsaturated soils can be solved. Considering the nonlinearity in the analysis, the iterative algorithm will be used in the solution process.

5.2 Coupling Equations for Seepage and Consolidation Analysis

In this chapter, the finite element formulations of the pore-water continuity equation and the equilibrium equations are derived respectively. The coupling equations of the combination of the continuity and equilibrium equations will be further adopted in the seepage-consolidation analysis for saturated-unsaturated soils. The coupling equations are as follows:

$$\begin{cases} [K]\{\Delta\delta\} - A_1[K_c]\{\bar{h}\} = \{\Delta R_F\} - A_1[K_c]\{\bar{h}_{n-1}\} \\ [K_c]^{T'}\{\Delta\delta\} + (\theta\Delta t[K_s] + [K_p])\{\bar{h}\} \\ = -\Delta t\{R_q\} + [K_p]\{\bar{h}_{n-1}\} - \Delta t(1-\theta)[K_s]_{n-1}\{\bar{h}_{n-1}\} \end{cases} \tag{5.41}$$

In Eq. (5.41), the unknowns in the pore-water continuity equation can be expressed in different forms, such as the form of the total head, the incremental form of the total head, the form of the pore-water pressure, and the incremental form of the pore-water pressure. Correspondingly, the coupling equations are also divided into different forms.

In the finite element program developed in this study, the total-head increment, Δh, is adopted as the unknown of the continuity equation. And the coupling equations are as follows:

$$\begin{cases} [K]\{\Delta\delta\} - A_1\gamma_w[K_c]\{\Delta h\} = \{\Delta R_F\} \\ -\gamma_w[K_c]^{T'}\{\Delta\delta\} - (\theta\Delta t\cdot[\dddot{K}_s] + [K_p]\gamma_w)\gamma_w\{\Delta h\} \\ = \Delta t\cdot\gamma_w\{R_q\} + \Delta t\cdot[\dddot{K}_s]\cdot\gamma_w\{h\}_{n-1} \end{cases} \tag{5.42}$$

Equation (5.42) can be further rewritten as follows:

$$\begin{bmatrix} [K] & -A_1\gamma_w[K_c] \\ -\gamma_w[K_c]^{T'} & -(\theta\Delta t\cdot[\dddot{K}_s]+[K_p]\gamma_w)\gamma_w \end{bmatrix}\begin{Bmatrix}\{\Delta\delta\}\\ \{\Delta h\}\end{Bmatrix}=\begin{Bmatrix}\{\Delta R_F\}\\ \Delta t\cdot\gamma_w(\{R_q\}+[\dddot{K}_s]\{h\}_{n-1})\end{Bmatrix} \tag{5.43}$$

where

$$\begin{aligned}
&[\dddot{K}_s]=[K_s]\gamma_w;\\
&[K]=\int_V [B]^T[\bar{D}][B]\mathrm{d}V;\\
&[K_c]=\int_V [B]^T\{M\}[\bar{N}]\mathrm{d}V;\\
&[K_c]^{T'}=\int_V \tilde{S}_r[\bar{N}]^T\{M\}^T[B]\mathrm{d}V;\\
&[K_s]=\int_V [B_s]^T[k][B_s]\mathrm{d}V;\\
&[K_p]=\int_V S_s[\bar{N}]^T[\bar{N}]\mathrm{d}V.
\end{aligned}$$

5.3 Initial and Boundary Conditions

In numerical analysis of seepage and consolidation, initial conditions and boundary conditions are the necessary preconditions for the solution. The initial condition is the starting point of the numerical analysis; the boundary condition can be regarded as the driving force for the changing of the stress, deformation, and seepage state.

For seepage-consolidation analysis of earth-rock dams, the state of soil before loading and impouding is generally taken as the initial state. The values of some basic variables of the initial conditions can be expressed as follows:

$$\begin{Bmatrix} u\\ \varepsilon\\ \sigma\\ u_a\\ u_w\\ S_r\\ e \end{Bmatrix}=\begin{Bmatrix} u_0\\ \varepsilon_0\\ \sigma_0\\ u_{a0}\\ u_{w0}\\ S_{r0}\\ e_0 \end{Bmatrix},\quad (x,y,z)\in\Omega \tag{5.44}$$

where Ω is the domain of the seepage-consolidation analysis; u is the displacement; ε is the strain; σ is the stress; u_a is the pore-air pressure; u_w is the pore-water pressure; S_r is the degree of saturation; e is the void ratio.

Typically, the values of the initial displacement and the initial pore-air pressure are zero, and the value of the initial pore-water pressure can be determined based on the construction status. During construction of earth-rock dams, the degree of saturation for the core material of clayey soil is about 85% ~ 90% after roller compaction, and the initial pore-water pressure can be then determined directly or indirectly. The initial values of other basic variables in seepage and consolidation analyses should also be determined based on the actual state of the dam materials.

For the seepage-consolidation analysis of earth-rock dams, the boundary conditions consist of the displacement boundary condition, the stress (or force) boundary condition, and the hydraulic boundary condition. And the hydraulic boundary condition also includes various types, such as the water head boundary, the flux boundary, and the seepage-face boundary.

(1) Displacement boundary condition

For this type of boundary condition, the displacement of the node is given. For instance, in the consolidation analysis of earth-rock dams, the contact zone between the dam body and the bedrock is usually set as a displacement boundary. And the displacement of this boundary is usually given as zero. The displacement boundary condition can be expressed as

$$\{\Delta u\} = \{\Delta \bar{u}\},\ (x, y, z) \in \Gamma^1 \tag{5.45}$$

where Γ is the boundary of the domain, Ω.

(2) Stress boundary condition

For earth-rockfill dams, the water of the reservoir will generate water pressure on the upstream surface of the dam. This kind of boundary belongs to the stress boundary. And the stress boundary condition can be express as

$$\left[\bar{M}\right]\{\Delta\sigma\} = \left\{\Delta\bar{F}\right\},\ (x, y, z) \in \Gamma^2 \tag{5.46}$$

where

$$\left[\bar{M}\right] = \begin{bmatrix} n_x & 0 & 0 & n_y & 0 & n_z \\ 0 & n_y & 0 & n_x & n_z & 0 \\ 0 & 0 & n_z & 0 & n_y & n_x \end{bmatrix} \tag{5.47}$$

(3) Head (or pore-water pressure) boundary condition

The head boundary condition is also known as the first-type boundary condition or the Dirichlet boundary condition. This boundary is the most commonly used boundary in seepage and consolidation analysis. It can be expressed as follows:

$$\{\Delta u_w\} = \{\Delta \bar{u}_w\},\ (x, y, z) \in \Gamma^3 \tag{5.48}$$

(4) Flux boundary condition

In seepage analysis, the flux boundary condition is also known as the second-type boundary condition or the Neumann boundary condition. In geotechnical engineering, the pore-water evaporation of soil, the infiltration of water into the soil, etc., are usually simulated by the flux boundary condition. This boundary condition can be expressed as follows:

$$lv_x + mv_y + nv_z = \bar{v}_n, \ (x, y, z) \in \Gamma^4 \tag{5.49}$$

where $\{N\} = \left\{ l \ m \ n \right\}^T$ is the normal direction of the boundary surface.

(5) Seepage-face boundary condition

In seepage and consolidation analysis of geotechnical engineerings, the head boundary and flux boundary can usually be determined directly. However, in some special cases, it is impossible to determine the hydraulic boundary directly prior to the analysis. The hydraulic boundary condition should be determined or adjusted during the analysis. For these cases, the boundary condition is named as the variable boundary condition or seepage-face boundary condition. For example, the seepage face of the downstream slope of a earth-rock dam after impounding, and the seepage face of the upstream surface of the dam after the reservoir water level plunged are all variable boundaries.

The location of the seepage face is unknown and can only be determined by iterative judgment. In seepage and consolidation analysis, the potential area of the seepage face needs to be set as the seepage-face boundary. The location of the seepage face changes with time and needs to be judged in each analysis stage (time step). The seepage face usually occurs in homogeneous earth dams without drainage. For earth-rock dams, due to good drainage of the dam shell material, there is generally no seepage face on the dam body in the seepage and consolidation analysis. However, for some special purposes, the core wall of an earth-rock dam should be analyzed separately, and then a seepage-face boundary should be set on the surface of the core wall.

The seepage-face boundary condition can be expressed as follows:

$$\{u_w\} = 0, \ (x, y, z) \in \Gamma^5 \tag{5.50}$$

where Γ^5 is the seepage face, which is determined through the iterative process.

5.4 Development of the Program for Multi-physics Coupling Analysis

5.4.1 Computation Procedure of the Finite Element Program

In this section, a finite element program for seepage-consolidation analyses of large-scale complex earth-rock dams is developed based on the multi-physics coupling consolidation analysis method. The computation flow of the program is shown in Fig. 5.1.

The FE program can perform multi-stage computations continuously to simulate the construction and impounding process of the earth-rock dams. The program includes a number of modules including the "file management", the "initialization of the calculation parameters", the "optimization of node numbers", the "processing of boundary conditions", the "calculation of degrees of freedom", the "load calculation", the "stiffness matrix assembly", the "solving of the linear equation system", the "parameter updating of constitutive models", the "output of results", and so on. In each computation (analysis) stage, the results of stress, deformation, water head and matric suction of the previous stage are taken as the initial conditions of the current stage. And according to the boundary conditions, the computation for unsaturated seepage and consolidation analysis of the current stage is performed.

5.4.2 Key Issues in the Finite Element Program Development

With the rapid development of computer technology, the FEM numerical simulation has become an indispensable research method in geotechnical engineering. Especially, with the construction of large-scale complex projects, the features and advantages of numerical simulation are more highlighted. Due to the high complexity of large-scale earth-rock dams, applying the FE program developed in this chapter to the coupling analysis of complex seepage and consolidation problems faces the following issues:

(1) Numerical stability;
(2) Computational accuracy;
(3) Computational efficiency.

The practicality and validity of the FE program are significantly affected by these problems. In view this, the following will give a brief description of the element types, the solution strategy for nonlinear problems, and the solution of large-scale sparse linear equation systems in the FE program.

Fig. 5.1 Flow chart of the finite element program

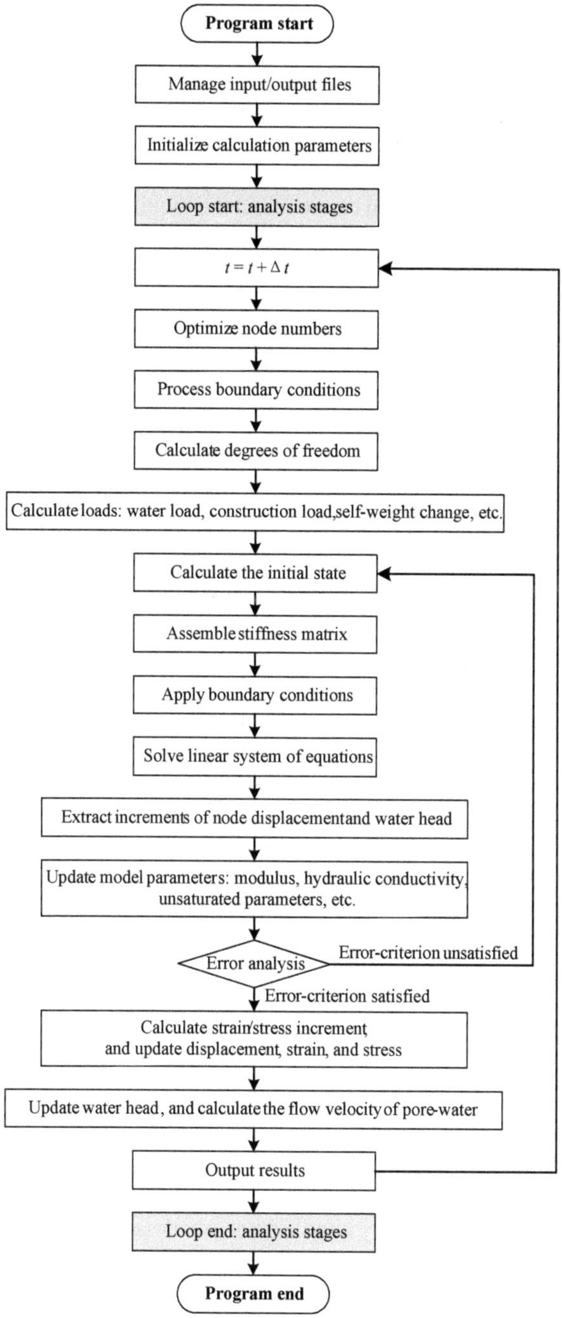

5.4.2.1 Element Types Used in the Seepage-Consolidation Analysis

In three-dimensional static FE analyses of earth-rock dams, the eight-node brick element is the most commonly used element type. Large-scale earth-rock dams have characteristics of the complex terrain and construction process and the numerous material zoning. Thus, when constructing the finite element model, a single type of hexahedral element cannot meet the requirements of adapting the mesh strictly to the terrain boundary, the construction process, and the material zoning, and there will be many degenerate elements in the finite element mesh.

In the numerical computation, these degenerate elements can still be treated as eight-node hexahedron elements and use the shape function and numerical integration scheme of the hexahedron element type. In this approach, the degenerate elements can be considered as special cases where some nodes of the hexahedron element overlap. This results in the over-estimation of the weights of the overlapped nodes in the degenerate elements, which will seriously affect the calculation accuracy of the program. Therefore, different types of isoparametric element are introduced in the FE program. For each type of degenerate elements, the corresponding more precise shape function and integration scheme will be adopted in the computation.

According to the research content of this dissertation and considering the computational precision and efficiency, the C_0 elements are selected as the basic elements in the finite element discretization. The C_0 elements include the brick (hexahedron) element, the wedge element, the tetrahedral element, and the pyramid element, as shown in Fig. 5.2. The eight-node hexahedral element is the most basic element in the three-dimensional numerical calculation of geotechnical engineering. For this element type, the numerical calculation is simple and accurate, but the adaptability to complex boundaries is poor. The introduction of other types of elements, i.e., the wedge element, the tetrahedral element, and the pyramid element, greatly enhances the ability of FE mesh to adapt to complex boundary conditions. The above types of elements will be used in combination when constructing a finite element mesh. The shape functions and integration schemes have been described in detail in some references and will not be repeated in this section (Lin and Li 1995; Zhu 2009; Dong et al. 2015).

5.4.2.2 Solution Strategy for Nonlinear Problems

Numerical stability is another problem that should be considered in the numerical analysis. In traditional consolidation analysis methods for saturated soils, the hydraulic boundary condition is greatly simplified, which avoids the nonlinear iteration in solving the pore-water continuity equation of the coupled consolidation analysis equations. The simplification of the hydraulic boundary condition makes the difficulty of solving the nonlinear problem significantly reduce and brings good computational stability. However, the computational precision is difficult to guarantee. In this study, in order to perform unsaturated coupling consolidation analysis of earth-rock dams, accurate distributions of water head in each analysis stage must be

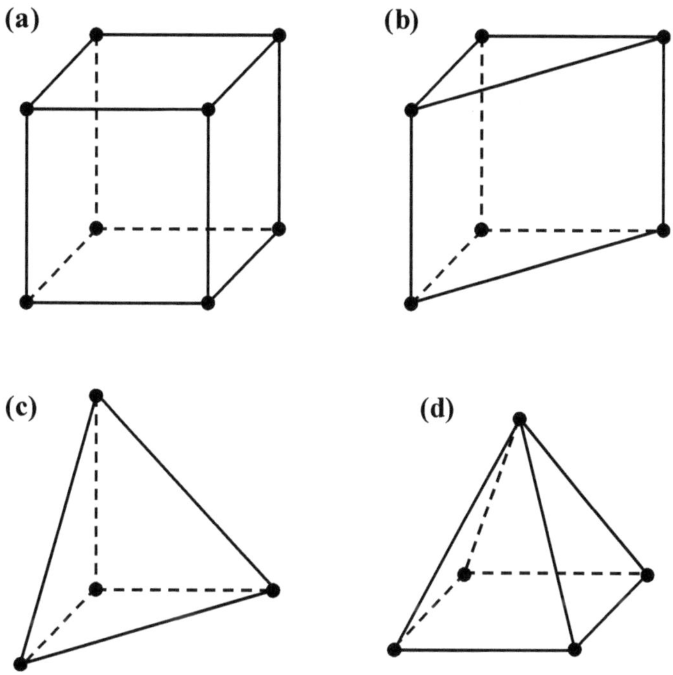

Fig. 5.2 Common C_0 element types in seepage-consolidation computation: **a** brick element; **b** wedge element; **c** tetrahedral element; **d** pyramid element

obtained. Hence, effective nonlinear iteration strategy should be adopted to solve the nonlinear problem.

Earth-rock dams usually contain multiple material zones. There are significant differences in the mechanical and hydraulic properties of the materials in different zones. In terms of mechanical properties, the deformation modulus and strength of the core material are quite different from those of the rockfill materials. The modulus of the dam shell materials is much greater than that of the clayey soil of core material. In terms of the hydraulic properties, the hydraulic conductivity of a saturated clayey soil is usually around 1×10^{-9} m/s, and it is significantly affected by the stress, deformation, and degree of saturation of the soil. The hydraulic conductivity of the dam shell materials is very large, usually up to 1×10^{-3} m/s. The huge differences and variations of the material properties make the coefficient matrix of the coupled equations of seepage-consolidation highly ill-conditioned, and the material parameters highly nonlinear. In other words, the consolidation analysis of unsaturated soils is much more complicated than the saturated consolidation analysis and more difficult to converge than the unsaturated unsteady seepage analysis.

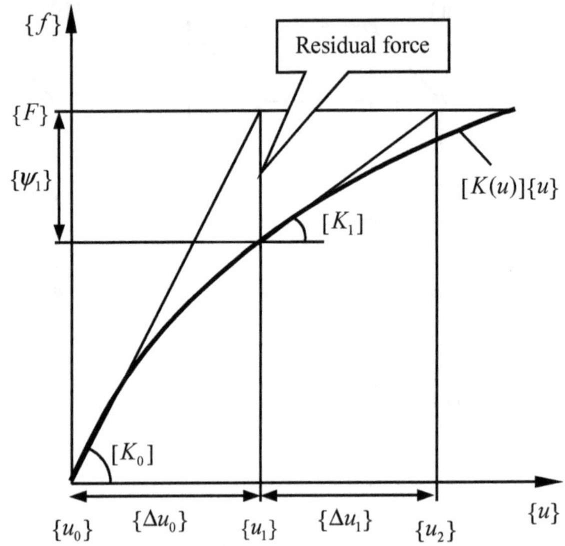

Fig. 5.3 Iteration process of the Newton-Raphson method

(1) Nonlinear iterative algorithm

By using the finite element method to solve the problems of material nonlinearity, there are two commonly used solutions, i.e., the incremental solution and the iterative solution. The incremental solution is easy to program, but its accuracy is relativity low. It shows obvious disadvantage in solving the strong nonlinear problems. On the contrary, the iterative solution has high precision, but its computational stability is relatively poor. In this study, due to the strong nonlinear relationship between the hydraulic conductivity and the water-head unknowns, an iterative solution (i.e., the Newton-Raphson method) is adopted to solve the nonlinear problem.

The Newton-Raphson method, also known as the tangent stiffness method, is one of the most commonly used methods for solving nonlinear problems. Figure 5.3 shows the iteration process of the Newton-Raphson method.

The FE equations for material nonlinear problems in geotechnical engineering can be expressed in the following form:

$$[K(u)]\{u\} = \{F\} \tag{5.51}$$

where $[K]$ is the global stiffness matrix; $\{u\}$ is the unknowns of node; $\{F\}$ is the nodal load.

Equation (5.51) can be written as follows:

$$\{\psi(u)\} = [K(u)]\{u\} - \{F\} = 0 \tag{5.52}$$

Assuming that the function $\{\psi\}$ is continuous and has the first-order derivative and retaining only the linear terms of the Taylor series expansions of $\{\psi\}$ at point $\{u_n\}$ yield

$$\{\psi\} = \{\psi_n\} + [K_T^n](\{u\} - \{u_n\}) = 0 \tag{5.53}$$

where $\{\psi_n\} = \{\psi(u_n)\}$, and $[K_T^n]$ is the tangent stiffness matrix at $\{u_n\}$.

Solving Eq. (5.53) yields an approximation of $\{u\}$ at the $(n+1)$th iteration, that is

$$\{u\} = \{u_n\} - [K_T^n]^{-1}\{\psi_n\} \tag{5.54}$$

Therefore, the solving process of the Newton-Raphson iterative method can be described as follows:

A. The initial value is given as $\{u_0\}$, and $n=0$.
B. Calculate the stiffness matrix $[K_T^n]$ at $\{u\} = \{u_n\}$.
C. Calculate the residual force:

$$\{\psi_n\} = \{\psi_n(u_n)\} = [K_T^n]\{u_n\} - \{F\} \tag{5.55}$$

D. Solve the system of equations:

$$\{\psi_n\} + [K_T^n]\{\Delta u_n\} = 0 \tag{5.56}$$

and yield,

$$\{\Delta u_n\} = -[K_T^n]^{-1}\{\psi_n\} \tag{5.57}$$

Thus, the value of $\{u\}$ for the next iteration is taken as:

$$\{u_{n+1}\} = \{u_n\} + \{\Delta u_n\} \tag{5.58}$$

E. Determine the convergence: if convergence, stop the iteration; if not convergence, then return back to step B and continue.

It can be seen that in the iteration process of the Newton-Raphson method, the stiffness matrix needs to be reassembled at every iteration step, and a new linear system of equations should be solved as well.

(2) Under-relaxation method for water-head iteration

In the seepage-consolidation analysis of unsaturated soils, the hydraulic conductivity of soil will change drastically when the degree of saturation changes, which is the main reason for the problems of numerical oscillation and non-convergence in the iterative solution process of nonlinear systems. For the unsaturated consolidation analysis and the unsaturated steady-unsteady seepage analysis, due to the strong nonlinear relationship between the hydraulic conductivity and the water head in the pore-water continuity equation, how to ensure the numerical stability becomes one of the difficulties in solving such problems. Especially for unreasonable parameter values of the constitutive models, the phenomena of iterative oscillation, slow-convergence, or even non-convergence will easily occur in the iteration process. These problems seriously affect the computational accuracy and efficiency of the program.

In seepage analysis using the numerical method, many scholars have studied on the problems encountered in the numerical solutions of pore-water continuity equation (Mehl 2006; Miller et al. 2006; Williams et al. 2000; Wu and Gao 1999). A common and effective solution is to apply the under-relaxation method to deal with the iterative unknowns (i.e., the water head or the pore-water pressure) in the nonlinear iteration process. The under-relaxation method has been widely used in commercial seepage analysis programs. Chen et al. (2012) compared a variety of under-relaxation methods and proposed a more practical and reliable mixed under-relaxation method.

On the basis of the mixed under-relation method, a four-terms under-relaxation method is adopted in the development of the FE program, as shown below:

$$\bar{h}_{n,m} = \alpha h_{n-1} + (1-\alpha)(\beta_1 h_{n,m} + \beta_2 h_{n,m-1} + \beta_3 h_{n,m-2} + \beta_4 h_{n,m-3}) \tag{5.59}$$

where $\bar{h}_{n,m}$ is the water head in the next iteration step for the current analysis stage; h_{n-1} is the water head for the previous analysis stage; $h_{n,m}$, $h_{n,m-1}$, $h_{n,m-2}$, and $h_{n,m-3}$ are the water head of the latest four iteration steps, respectively; and α and β are the under-relaxation parameters.

5.4.2.3 Solutions of the Large-Scale Sparse Linear Systems

The numerical solution of seepage and consolidation in geotechnical engineering is finally achieved by solving a series of linear equation systems. The linear equation systems can be expressed as follows:

$$\boldsymbol{Ax} = \boldsymbol{b} \tag{5.60}$$

where $\boldsymbol{A}$ is a sparse matrix of known coefficients in which most of the elements are zero; $\boldsymbol{b}$ is a column vector of known coefficients; $\boldsymbol{x}$ is the column vector of unknowns.

The solution of linear systems of equations is the most time-consuming part in a large-scale finite element analysis. The efficiency of solving the linear systems determines whether the finite element program is competent for large-scale engineering applications or not. With the development of the solution methods for linear equation systems, a lot of the efficient methods have been successfully applied to the numerical computation of geotechnical engineering.

Numerical methods for solving linear systems of equations can generally be divided into two classes, i.e., the direct methods and the iterative methods. In the finite element numerical analysis of seepage-consolidation, the direct methods (e.g., the decomposition methods) have been widely used in solving systems of linear equations. The direct methods are stable and easy to program and can yield the exact solution within a finite number of steps. For small-scale problems, the direct methods have obvious advantages. With the construction of large-scale geotechnical engineerings, the demand for FE computations is increasing, and the direct methods show great limitations. On the one hand, the storage schemes of the coefficient matrix are relatively backward; on the other hand, the computational efficiency of the solution methods is relatively low, which cannot meet the requirements of large-scale engineering problems. In view of the deficiencies of the traditional direct method, more efficient direct methods for sparse linear systems have been developed in recent years. Although the performance of the direct methods has been greatly improved, it is still insufficient for problems of super-large scale (Dong 2015). For large-scale sparse linear systems of equations, the Krylov subspace iterative methods combined with effective preprocessing methods show significant advantages over the direct methods (Chen and Phoon 2012; Zheng and Zhang 2007; Chen et al. 2006; Phoon et al. 2002). Currently, the application of preconditioned iterative methods in the numerical computations of large-scale problems is still rare (Chen et al. 2013; Liu et al. 2005; Zhang et al. 2004).

The research group of the author has studied numerous preconditioned iterative methods that are used to solve the large-scale symmetric-asymmetric linear systems. Many of the methods have been successfully applied to the numerical solutions of practical engineerings. The iterative methods include the symmetric quasi-minimal residual (SQMR) method, the minimal residual (MINRES) method, the conjugate gradient (CG) method, the stabilized Bi-conjugate gradient (Bi-CGSTAB), the quasi-minimal residual (QMR) method, the conjugate gradients-squared (CGS) method, the generalized minimum residual (GMRES) method, the induced dimension reduction (IDRs) method (Sonneveld and van Gijzen 2008), and so on. And the preconditioning methods include the Jacobi method, the symmetric successive over-relaxation (SSOR) method, the modified block SSOR (MBSSOR) method, and so on.

Table 5.1 shows the scope of application of the preconditioning methods and the iterative methods. All the iterative methods shown in the table can be used for a symmetric matrix. For the general asymmetric matrix, the Bi-CGSTAB method, the QMR method, the CGS method, the GMRES method, and the IDRs method are applicative.

Table 5.1 Application scope of the preconditioned iterative solutions (Dong 2015)

Method	SQMR	MINRES	CG	Bi-CGSTAB	QMR	CGS	GMRES	IDRs
Symmetric matrix	√	√	√	√	√	√	√	√
Asymmetric matrix				√	√	√	√	√
(G)Jacobi	√	√	√	√	√	√	√	√
(M)SSOR	√	√	√	√	√			
MBSSOR	√							

5.5 Influence of Hydraulic Characteristics of Soil on the Seepage Field

In traditional saturated consolidation analysis, the soil is assumed to be completely saturated, and a constant saturated hydraulic conductivity is taken as the hydraulic conductivity of the soil. This is a significant simplification in the analysis. In numerical analysis of seepage-consolidation for unsaturated soils, the values of the hydraulic characteristic parameters directly affect the results of seepage in the soils. Without considering the unsaturated hydraulic characteristics of the soil will yield incorrect results. For unsaturated soils, to describe the unsaturated flow phenomena exactly, two constitutive relationships are needed, namely, the hydraulic conductivity and the soil-water characteristic curve. Thus, from these two aspects, the influence of the hydraulic characteristics of soil on the unsaturated flows are studied using a simple example.

A soil column, as shown in Fig. 5.4, is selected for seepage-consolidation analysis to study the distribution of the seepage field. This soil model contains a total of two material zones. The material of the left zone is a clayey soil, and the material of the right zone is a sandy soil. The hydraulic boundary condition of the model for seepage-consolidation analysis is shown in the figure. This model has a higher water level on the left boundary and a lower water level on the right boundary, resulting in the flows from left to right.

Firstly, the distributions of seepage field in the following three cases are analyzed to study the influence of the unsaturated hydraulic conductivity on the results of seepage analysis:

(1) The hydraulic conductivities of the two soils in the soil model are calculated by using the mathematical model of unsaturated hydraulic conductivity in Eq. (4.67);
(2) The hydraulic conductivity of the clayey soil takes the saturated hydraulic conductivity (ignoring the influence of stress, deformation, and degree of saturation on its permeability), and the hydraulic conductivity of the sandy soil is calculated by using Eq. (4.67);

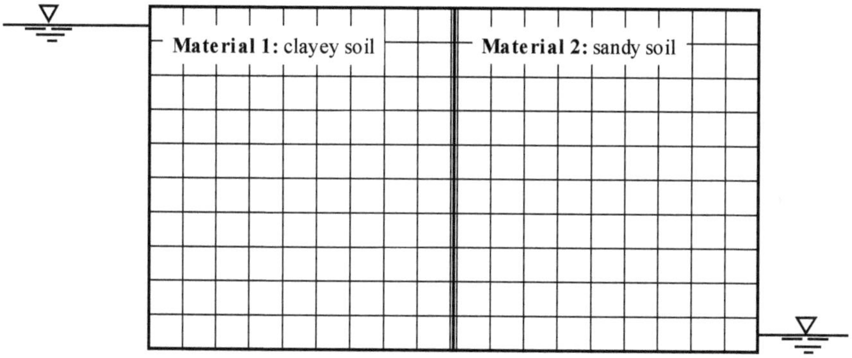

Fig. 5.4 Soil column model for seepage analysis

Table 5.2 Values of the hydraulic characteristic parameters

Material	k_s (m/s)	Parameters of van Genuchten model			
		α (m^{-1})	n	θ_s	θ_r
Clayey soil	1×10^{-9}	0.8	1.09	0.38	0.068
Sandy soil	1×10^{-4}	14.5	2.68	0.43	0.045

(3) In the analysis, the saturated hydraulic conductivities of the two soils are taken as their hydraulic conductivities.

For the three cases, the parameters of hydraulic characteristics used in the analyses are shown in Table 5.2. The saturated hydraulic conductivity is assumed to be constant, and the soil-water characteristic curve parameters are taken from the relevant literature (Carsel and Parrish 1988). Figure 5.5 plots the phreatic lines in the soil column for the three cases when the seepage-consolidation reaches a steady state, and the three phreatic lines correspond to the three cases, respectively. Here, a phreatic line refers to the zero contour of pore-water pressure.

As can be seen from Fig. 5.5, line #1 is at the highest position among the three. This phreatic line gradually decreases in the direction of seepage and outflow from the clay at about 1/2 height of the clay-sand interface.

The positions of phreatic lines #2 and #3 are relatively low, and there is no seepage face at the clay-sand interface. It is obvious that the distributions of lines #2 and #3 are incorrect and will not exist in actual situations. In these two cases, the saturated and unsaturated regions of the clayey soil are assumed to have the same permeability, and their resistances to water flowing are also the same so that the soil does not need to conduct water by increasing the degree of saturation. Due to the fact that the permeability of soil decreases significantly in unsaturated regions, the seepage flux in the unsaturated region is apparently overestimated. And thus, the left ends of line #2 and line #3 are not perpendicular to the boundary of the upstream face.

In saturated seepage analysis, the phreatic line is the upper boundary of the seepage area. Thus, it is also a flow line. For unsaturated seepage problems, the phreatic

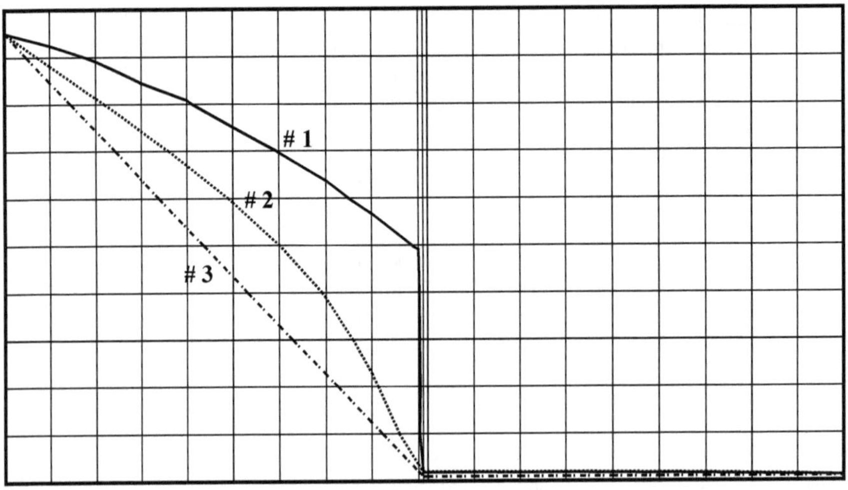

Fig. 5.5 Distribution of the phreatic lines in different cases

line is the dividing line between the saturated and unsaturated regions of soil. The phreatic line is not a strict flow line because water can still flow across it. However, the permeability of soil decreases significantly in the unsaturated region, and the difficulty for the water to flow from the saturated region into the unsaturated region greatly increases. Hence, the phreatic line is approximately parallel to the adjacent flow lines. In addition, the boundary of the upstream face is an equipotential line, and the flow lines should be perpendicular to the upstream face. Thus, the phreatic line in unsaturated seepage should be approximately perpendicular to the upstream face. When the unsaturated permeability of a soil is assumed to be the same as the saturated permeability, more water will flow through the phreatic line. Hence, the phreatic line will no longer parallel to the adjacent flow lines.

In the traditional saturated consolidation analysis, the soil is fully saturated, and its permeability is assumed to be constant. Under this simplification, the pore-water continuity equation is usually solved without using the nonlinear iterative method. In order to obtain approximately reasonable results, the hydraulic boundaries need to be treated properly. For a typical earth-rock dam shown in Fig. 5.6, the hydraulic conductivity of the core wall is much lower than that of the dam shell material. In traditional saturated consolidation analysis, in order to simplify the calculation, only the earth core wall is selected as the object of the water-flow analysis. The hydraulic boundaries of the core wall can be divided into three categories, as shown in the figure. And for the second boundary, the pressure head is assumed to be zero. Obviously, the above simplification will induce relatively big simulation errors.

Then, seepage analyses with different parameter sets of the soil-water characteristic curve are performed on the soil column model shown in Fig. 5.4. In the analyses, the saturated hydraulic conductivities of material 1 and material 2 take the values of k_s for clayey soil and sandy soil in Table 5.2, respectively. And the SWCC parameters

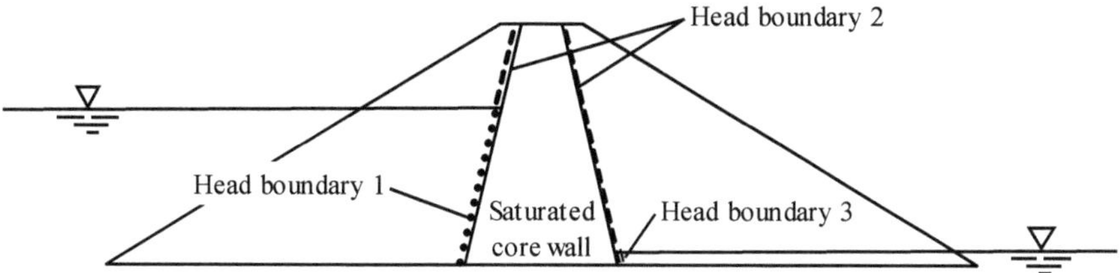

Fig. 5.6 Simplification of the water head boundary in traditional consolidation analysis

Table 5.3 Parameters of soil-water characteristic curve (van Genuchten model)

Parameter set	Material zone 1		Material zone 2	
	α (m^{-1})	n	α (m^{-1})	n
1	0.1	1.10	0.1	1.10
2	0.1	1.10	2.0	1.50
3	2.0	1.50	2.0	1.50
4	2.0	1.50	10.0	2.20

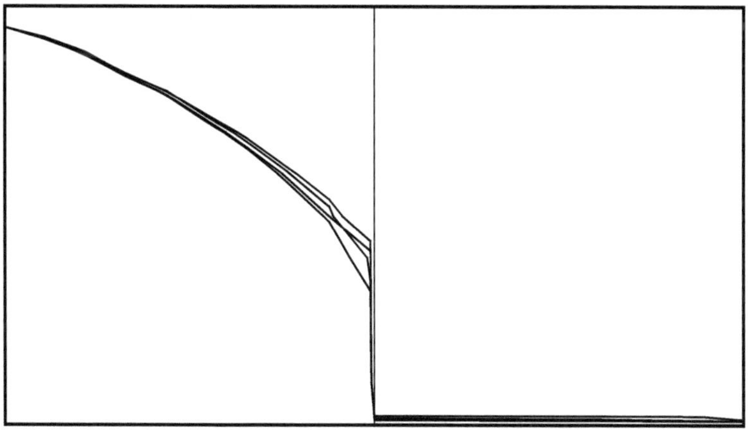

Fig. 5.7 Distribution of the phreatic lines for different SWCC parameters

of the two material zones take the values shown in Table 5.3. There are a total of four sets of parameters in the table. And the range of the SWCC parameters basically covers the parameter values from clayey soils to coarse-grained soils. The parameter values of set 1 are relatively close to those of the clayey soil, and the parameter values of set 4 are relatively close to those of the coarse-grained soils.

Numerical simulations of seepage are performed by using the four sets of SWCC parameters. The phreatic lines of the four sets of parameters are plotted in Fig. 5.7. And Fig. 5.8 shows the distributions of the total head for the parameter sets.

It can be seen from the results that the influence of the SWCC parameters on the distribution of phreatic line is not significant. That is, the parameters have an

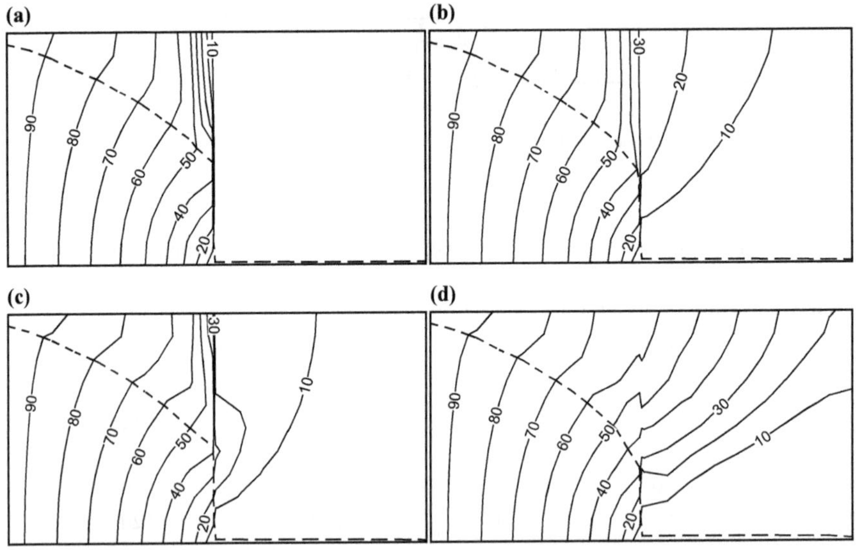

Fig. 5.8 Distributions of the total head for different SWCC parameters (m): **a** parameter set 1; **b** parameter set 2; **c** parameter set 3; **d** parameter set 4

inapparent effect on the bound of the saturated area and on the distribution of water head within the saturated area. However, when analyzing the mechanical behaviors of soils, the unsaturated areas also are the focus of attention, and whether the SWCC parameters can reflect the true water-holding characteristic of the soil becomes particularly important.

The analyses in this section have only been aimed at the final state when the seepage and consolidation reach a steady state. Beyond that, the hydraulic characteristics of soil also have significant influences on the process of seepage and consolidation. Therefore, reasonable hydraulic parameters are very important in the unsaturated seepage-consolidation analysis.

5.6 Unsaturated Seepage-Consolidation Analysis for Typical Examples

In order to prove the validity and stability of the multi-physics coupling analysis program developed in this chapter, two ideal two-dimensional earth-rock dams are adopted as typical examples to perform numerical simulations of unsaturated seepage and consolidation. The examples and the analysis results are as follows.

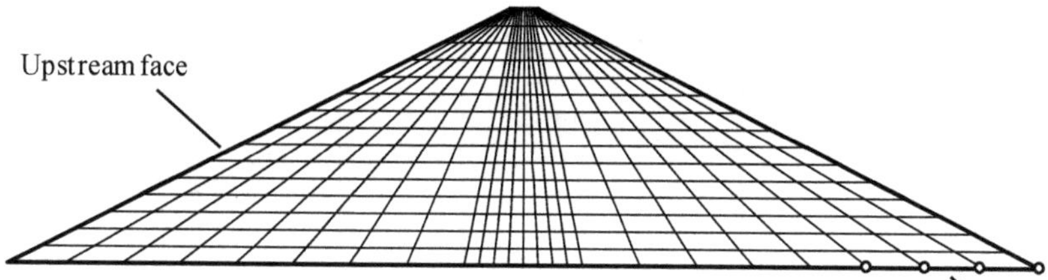

Fig. 5.9 Two-dimensional FE model of the homogeneous earth dam

Table 5.4 Parameters of Duncan and Chang's EB model

K	n	R_f	K_b	m	φ_0 (°)	$\Delta\varphi$ (°)
420	0.50	0.78	300	0.25	39.5	9.7

Table 5.5 Relevant unsaturated parameters

m_2	s_e (kPa)	C_h	ξ
0.3	10.0	0.02	1.0

Table 5.6 Parameters of hydraulic characteristics

α (m^{-1})	β	θ_r	θ_s	a	b	c (m/s)
0.005	0.56	0.10	0.38	20.8	−1.0	1.3×10^{-9}

5.6.1 Example 1: The Homogeneous Earth Dam

5.6.1.1 Simulation Model and Parameters

Example 1 gives a homogeneous earth dam on an impervious foundation. The height of the dam is 90 m, and a horizontal drainage blanket is installed over the portion of the downstream foundation. The FE program developed in this study is used to simulate the construction and impounding process of the dam. The dam body is constructed by stratified filling. After construction, the upstream water level slowly rises to the normal water level. Figure 5.9 shows the finite element model of this dam, which contains 800 nodes, and 360 elements. The boundary condition of the FE model includes the displacement constraints at the bottom of the dam and the head boundaries on the upstream and downstream sides. The main parameters used in this example are shown in Tables 5.4, 5.5 and 5.6.

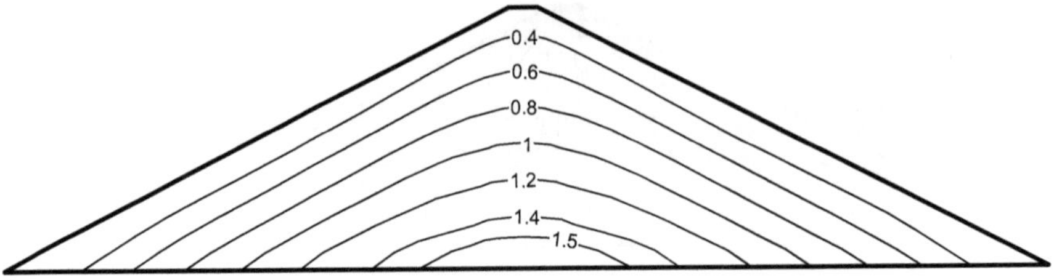

Fig. 5.10 Distribution of the major principal stress at completion (MPa)

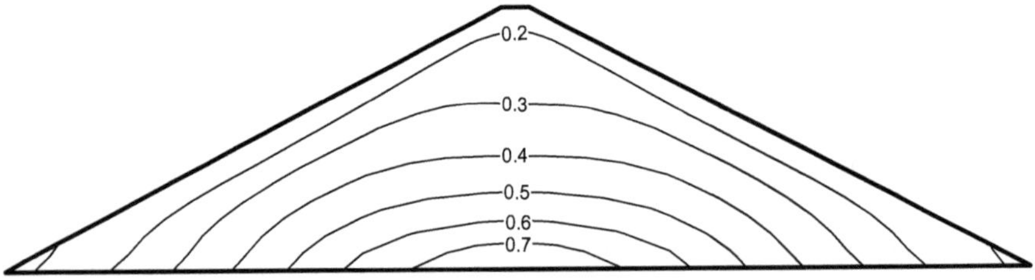

Fig. 5.11 Distribution of the minor principal stress at completion (MPa)

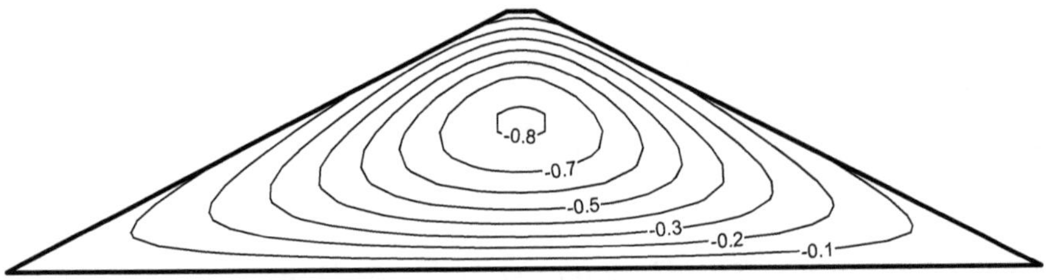

Fig. 5.12 Distribution of the settlement at completion (m)

5.6.1.2 Results and Analyses

The above parameters are adopted to analyze the unsaturated seepage and consolidation characteristics of the homogeneous earth dam. Figures 5.10, 5.11, 5.12 and 5.13 show the distributions of stress and deformation in the dam at the completion of construction (before impounding). It can be seen that the maximum and minimum principal stress in the dam are about 1.5 and 0.7 MPa, respectively. The stress on the surface of the dam is the smallest, and it gradually increases from the surface to the interior of the dam. The maximum vertical settlement within the dam is about 0.8 m, which is located in the center of the dam. And the maximum horizontal displacement in the dam is about 0.25 m, showing a symmetrical distribution.

Figures 5.14, 5.15, 5.16, 5.17 and 5.18 show the distributions of the stress, the pressure head, the flow velocity of pore water, etc., in the dam after impounding. It can be seen from the stress distributions that the effective stress in the dam is significantly decreased after impounding, especially for the effective stress in the

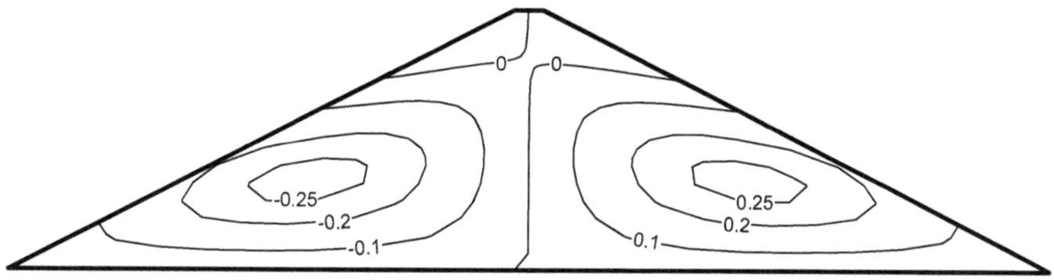

Fig. 5.13 Distribution of the horizontal displacement at completion (m)

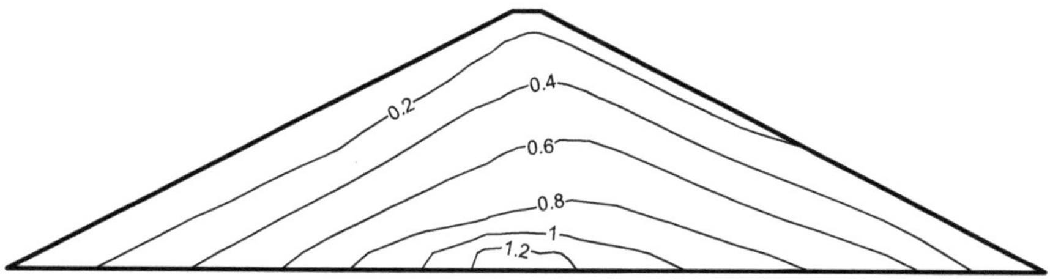

Fig. 5.14 Distribution of the major principal stress after impounding (MPa)

upstream side of the dam. Figure 5.16 shows the distribution of pressure head. As shown in the figure, the zero contour line of pressure head is the phreatic line in the dam, the position of which is effectively reduced by the downstream horizontal drainage blanket. In this example, it is assumed that the pore-air in the dam is fully drained. Hence, the phreatic line is the boundary between the saturated area and the unsaturated area of the dam. The soil below the phreatic line is completely saturated, with positive pore-water pressure. And the soil above the phreatic line is unsaturated, with negative pore-water pressure. Figure 5.17 shows the distributions of the water head and the flow velocity in the dam. The water head decreases gradually along the seepage path. The arrows in the figure show the direction and the magnitude of the flow velocity. It can be seen that water seepage mainly occurs in the saturated area of the dam, although seepage also occurs in the unsaturated area above of the phreatic line. This kind of distribution is consistent with the results obtained by unsaturated seepage analysis. Figure 5.18 shows the distribution of the matric suction in the dam after impounding. The contour line of matric suction reflects the distribution of the negative pore-water pressure in the dam. And the matric suction is zero in the saturated area below the phreatic line.

Comparing the multi-physics coupling analysis program developed in this chapter with the traditional programs for saturated consolidation analysis, it is obvious that the results of the phreatic line, the flow velocity, etc., cannot be obtained in the traditional consolidation analysis. This is one of the significant advantages of the multi-physics coupling analysis program.

The above are the main results of the seepage and consolidation analysis of an ideal homogeneous dam using the FE program develop in this study. In order to reveal the effects of the unsaturated conditions, the homogeneous dam of this example is

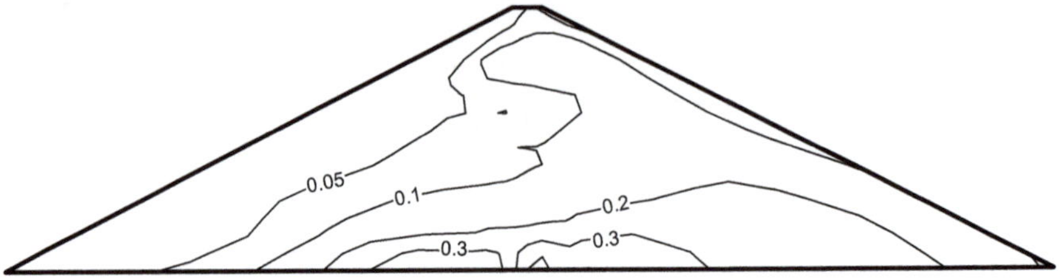

Fig. 5.15 Distribution of the minor principal stress after impounding (MPa)

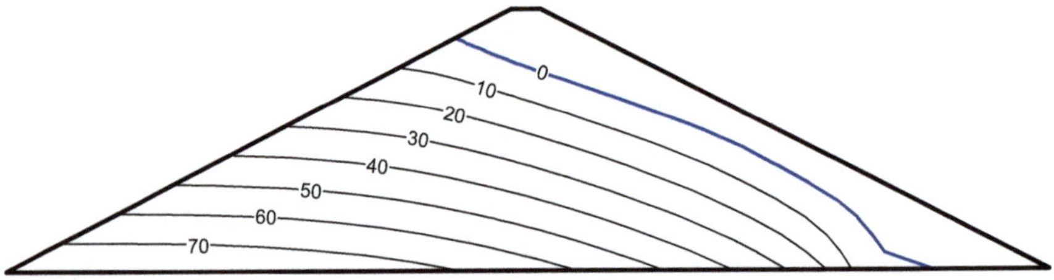

Fig. 5.16 Distribution of the pressure head after impounding (m)

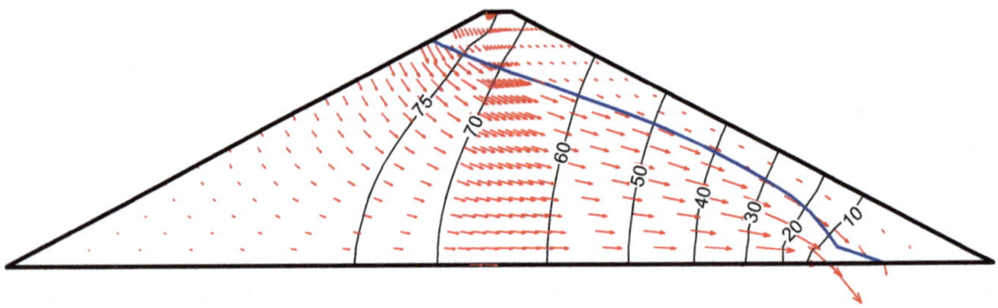

Fig. 5.17 Distributions of the total head (m) and the flow velocity after impounding

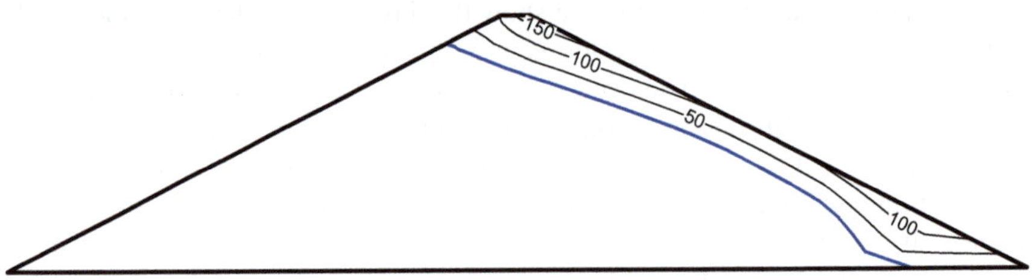

Fig. 5.18 Distribution of the matric suction after impounding (kPa)

also analyzed using a saturated consolidation analysis program. And Figs. 5.19, 5.20, 5.21 and 5.22 show the distribution of the stress and deformation in the dam at completion.

Comparing the stress results in Figs. 5.19 and 5.20 with those in Figs. 5.10 and 5.11 respectively, it can be seen that there are significant differences between the

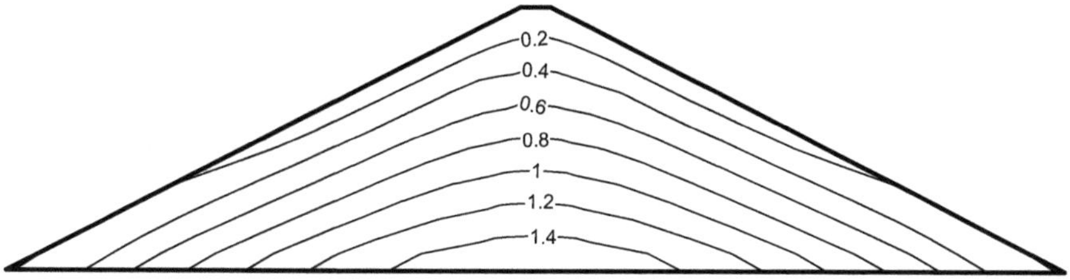

Fig. 5.19 Distribution of the major principal stress (MPa)

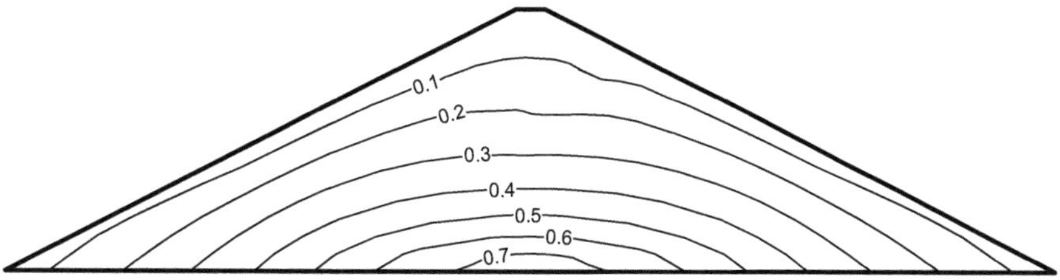

Fig. 5.20 Distribution of the minor principal stress (MPa)

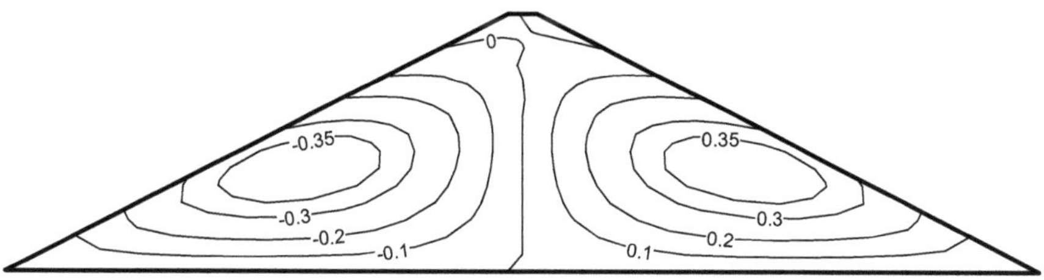

Fig. 5.21 Distribution of the horizontal displacement (m)

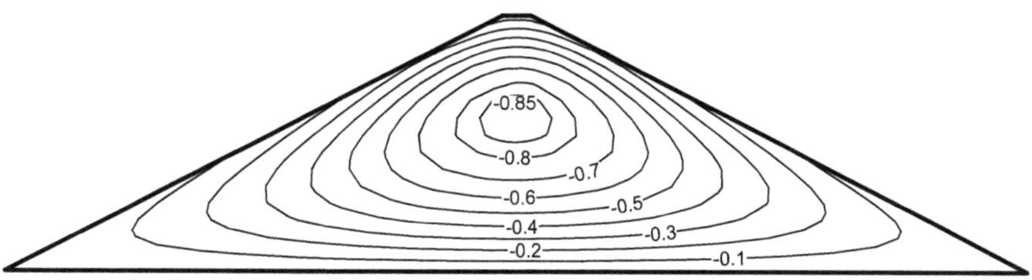

Fig. 5.22 Distribution of the settlement (m)

effective stress values calculated by the two programs, especially for the upper areas with a relatively low degree of saturation. The effective stress calculated by the unsaturated program developed in this study is obviously greater than that calculated by the traditional saturated consolidation program.

Comparing the displacement results of the dam in Figs. 5.21 and 5.22 with the results in Figs. 5.12 and 5.13 respectively, there are also significant differences. In the results of the saturated program, the maximum value of the vertical displacement is about 0.88 m, and the maximum value of the horizontal displacement reaches 0.39 m, which are larger than those calculated by the unsaturated program.

The above differences can be explained by the mechanism as follows.

In the consolidation analysis using the unsaturated theory, the soil of the dam before impounding will be unsaturated. The effective stress of the soil is calculated by Eq. (5.61):

$$\sigma' = \sigma^* + \chi s \tag{5.61}$$

When the soil is unsaturated, there will be matric suction in the soil. The matric suction acts on the soil and has a significant contribution to the effective stress of the soil. However, this contribution is neglected in the analysis using the saturated consolidation theory. Therefore, the effective stress calculated by the unsaturated program is greater than that calculated by the saturated program.

In addition, the effective stress in the soil significantly affects the strength and deformation characteristics of the soil. The deformation modulus of soil will increase with the increase of the effective stress. Therefore, the displacement calculated by the unsaturated program is slightly smaller than that of the results calculated by the saturated program. It can be seen that the results of the unsaturated program are more comprehensive and reasonable than those of the saturated program.

5.6.2 *Example 2: The Earth-Rockfill Dam*

5.6.2.1 Simulation Model and Parameters

Example 2 gives an ideal earth-rockfill dam on an impervious foundation. The finite element model of this dam is shown in Fig. 5.23. There are three types of dam materials in total, including the upstream rockfill, the downstream rockfill, and the core material. The initial state of the core material is assumed to be saturated. The boundary condition of the FE model includes the displacement constraints at the bottom of the dam and the head boundaries on the upstream and downstream sides. The main parameters used in this example are shown in Tables 5.7, 5.8 and 5.9.

5.6.2.2 Results and Analyses

The Unsaturated seepage and consolidation analysis is carried out using the above FE model and parameters. Figures 5.24, 5.25, 5.26, 5.27 and 5.28 show the distributions of the stress, the displacement, and the pore-water pressure in the dam at the completion of construction.

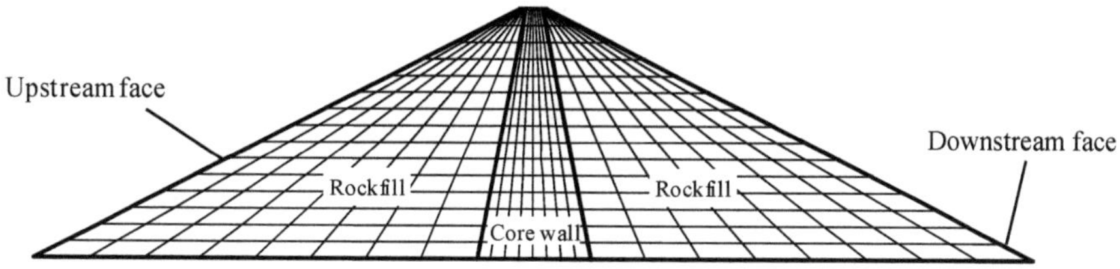

Fig. 5.23 Material zoning and boundary condition of the FE model

Table 5.7 Parameters of Duncan and Chang's EB model

Material	K	n	R_f	K_b	m	φ_0 (°)	$\Delta\varphi$ (°)
Upstream rockfill	1491	0.241	0.719	600	0.110	54.37	10.47
Core wall	388	0.311	0.755	230	0.145	39.47	9.72
Downstream rockfill	1491	0.241	0.719	600	0.110	54.37	10.47

Table 5.8 Parameters of hydraulic characteristics

Material	α (m^{-1})	β	θ_r	θ_s	a	b	c (m/s)	k_s (m/s)
Upstream rockfill	0.05	3.15	0.05	0.38	\	\	\	1×10^{-4}
Core wall	0.005	0.56	0.10	0.38	20.8	−1.0	1.4×10^{-14}	\
Downstream rockfill	0.05	3.15	0.05	0.38	\	\	\	1×10^{-4}

Table 5.9 Relevant unsaturated parameters

Material	m_2	s_e (kPa)	C_h	ξ
Core wall	0.3	10.0	0.02	1.0

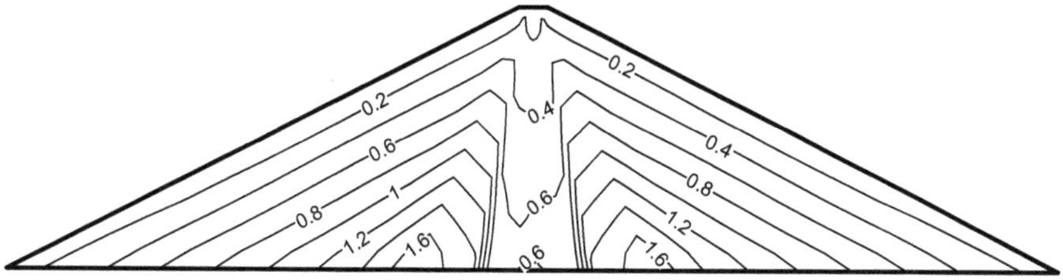

Fig. 5.24 Distribution of the major principal stress at completion (MPa)

From the stress distributions, it can be seen that the effective stress in the core wall is significantly smaller than that in the rockfill, which is caused by the uncoordinated deformations between the core material and the rockfill material. In the analysis of this example, the matric suction in the upstream and downstream rockfills is neglected. And for the clayey soil of the core material, the effect of the matric suction

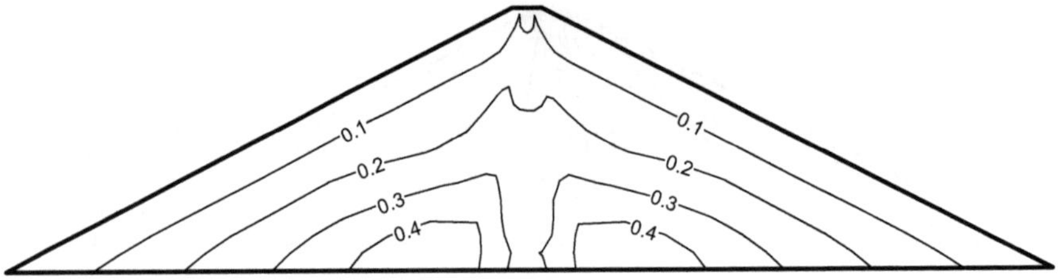

Fig. 5.25 Distribution of the minor principal stress at completion (MPa)

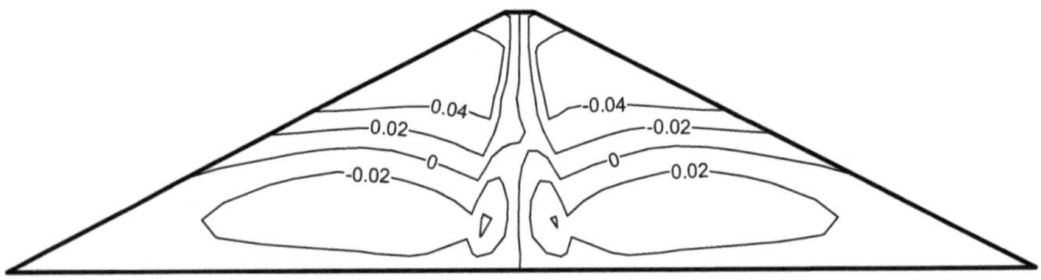

Fig. 5.26 Distribution of the horizontal displacement at completion (m)

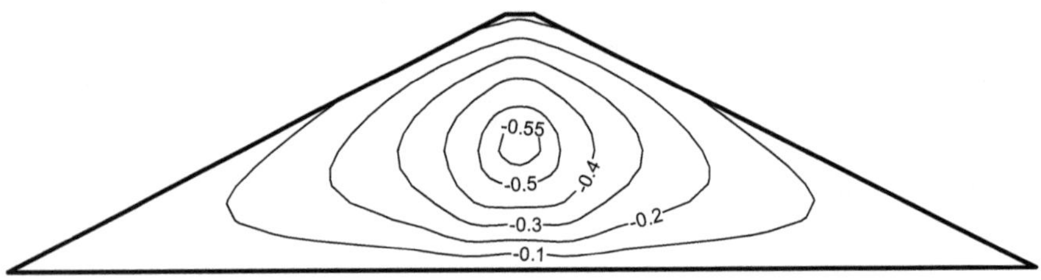

Fig. 5.27 Distribution of the settlement at completion (m)

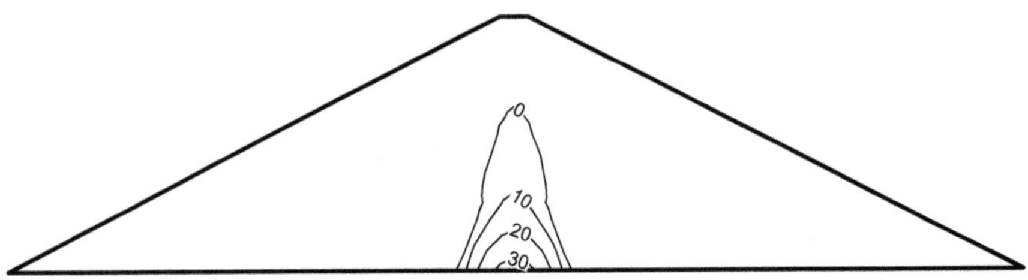

Fig. 5.28 Distribution of the pressure head in the core wall at completion (m)

is taken into account in the calculation of the effective stress. From the results of the displacement in the dam at completion, the horizontal displacement in the dam shows a symmetrical distribution, and the maximum settlement is located in the center of the dam.

Figure 5.28 shows the distribution of the pore-water pressure in the core wall at the completion of the dam. It can be seen that the pore-water pressure in the middle

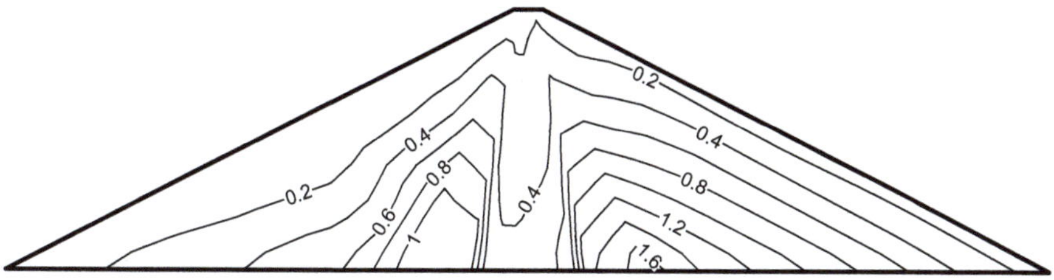

Fig. 5.29 Distribution of the major principal stress after impounding (MPa)

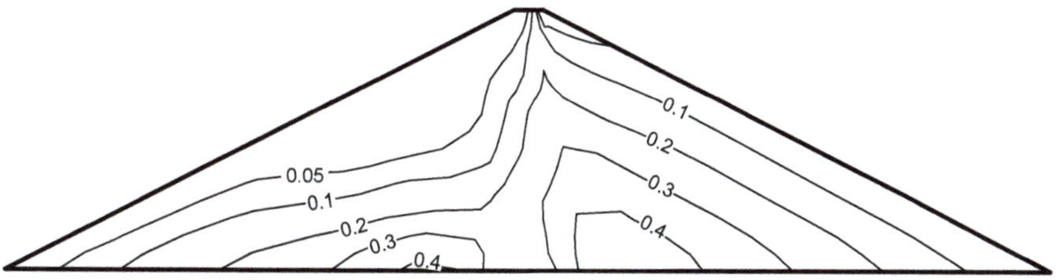

Fig. 5.30 Distribution of the minor principal stress after impounding (MPa)

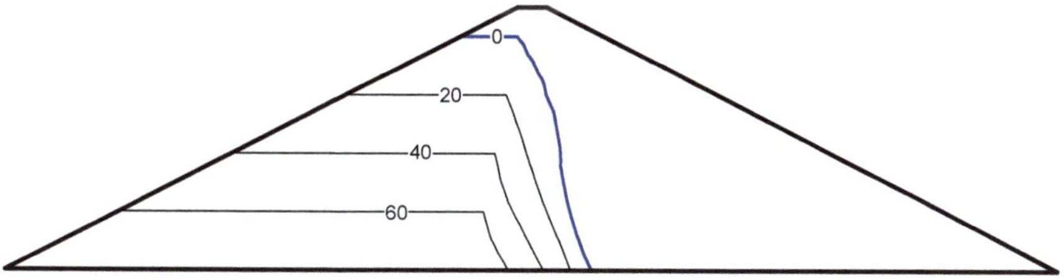

Fig. 5.31 Distribution of the pressure head after impounding (m)

and lower part of the core wall is positive and reaches the maximum value at the bottom of core wall. Due to a high speed of construction and low permeability of the core wall, the excess-pore water pressure at completion caused by the construction load has not completely dissipated. The existence of the pore-water pressure in the core wall further reduces the effective stress, which is another reason why the stress in the core wall is much smaller than that in the rockfill.

Figures 5.29, 5.30, 5.31 and 5.32 show the distributions of the stress and the water head in the dam after reservoir impounding. From the stress distribution in the figures, the effective stress in the upstream side of the dam is significantly reduced due to the impoundment, and the stress in the downstream rockfill is basically unchanged. From the distribution of the water head, the hydraulic gradient in the dam mainly occurs in the core wall, which reflects an good imperviousness of the core wall.

On the whole, the seepage-consolidation analysis results of the two ideal earth-rock dam examples using the coupling analysis program developed in this chapter are reasonable.

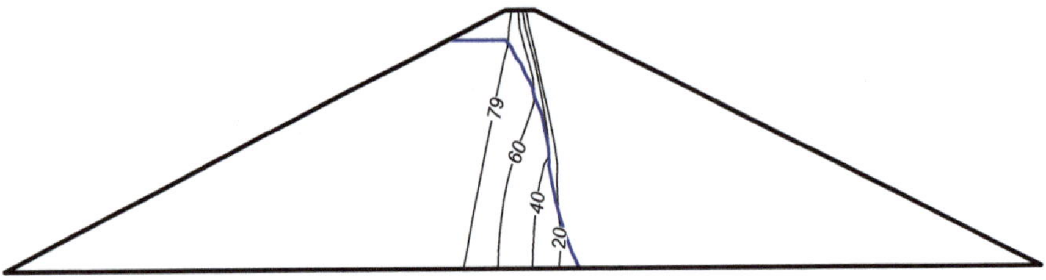

Fig. 5.32 Distribution of the total head after impounding (m)

5.7 Summary

In this chapter, the numerical method of multi-physics coupling seepage and consolidation analysis for unsaturated soils is studied. On the basis of the multi-physics coupling analysis method proposed in this study, combined with the finite element method for programming, the numerical solution of unsaturated seepage-consolidation for complex engineering problems is achieved. Then, two ideal examples of earth-rock dams are given to validate and analyze the theory and program of multi-physics coupling analysis. The main results and conclusions are as follows:

(1) Based on the simplified consolidation theory for unsaturated soils and the multi-physics coupling analysis method, the finite element equations for unsaturated seepage and consolidation analysis are derived by using the Galerkin method, and then a finite element program is developed.
(2) Multiple element types, the iterative method for nonlinear problems, the under-relaxation method, and the iterative algorithms for large-scale linear systems of equations are introduced into the FE program, which effectively improves the computational accuracy and efficiency and the numerical stability of the program.
(3) By using the FE program developed in this chapter, the unsaturated seepage consolidation analyses of two earth-rock dam examples are carried out, and reasonable numerical results are obtained. Comparing the results of the FE program with those of a traditional saturated consolidation program, it can be seen that the matric suction in the unsaturated soil has significant effects on the stress and deformation characteristics in unsaturated consolidation analyses.
(4) Compared with the traditional saturated analysis program, the multi-physics coupling analysis program developed in this study has significant advantages. For example, the newly developed program can perform unsaturated seepage and consolidation analyses; it considers the changes of soil permeability with the stress, deformation, and physical state; the effects of the matric suction on the stress, deformation, and strength are taken into account; and by using a nonlinear iterative solution in the program, accurate seepage-field results (e.g., distributions of the water head and the phreatic line) can be obtained.

References

Biot MA (1941) General theory of three-dimensional consolidation. J Appl Phys 12(2):155–164
Biot MA (1955) Theory of elasticity and consolidation for a porous anisotropic solid. J Appl Phys 26(2):182–185
Booker JR, Small JC (1975) An investigation of the stability of numerical solutions of Biot's equations of consolidation. Int J Solids Struct 11(7–8):907–917
Carsel RF, Parrish RS (1988) Developing joint probability distributions of soil water retention characteristics. Water Resour Res 24(5):755–769
Chandrupatla TR, Belegundu AD, Ramesh T, Ray C (2002) Introduction to finite elements in engineering, vol 2. Prentice Hall Upper Saddle River, NJ
Chen X, Liu JK, Li X, Tian YH, Wang YN (2013) A high-performance preconditioner for iterative solutions of geotechnical applications. Rock and Soil Mechanics 34(S2):S2
Chen X, Phoon KK (2012) Applications of symmetric and nonsymmetric MSSOR preconditioners to large-scale Biot's consolidation problems with nonassociated plasticity. Journal of Applied Mathematics 2012
Chen X, Toh KC, Phoon KK (2006) A modified SSOR preconditioner for sparse symmetric indefinite linear systems of equations. Int J Numer Meth Eng 65(6):785–807
Chen X, Yu YZ, Cheng YG (2012) Under-relaxation methods for numerical solution of Richards equation of variably saturated flow. Rock and Soil Mechanics 33(S1):237–243
Dong W (2015) Elasto-plastic fluid-solid coupling analysis of seismic response of high core-wall rockfill dam. Doctroral dissertation, Tsinghua University, Beijing
Dong WX, Wang XN, Wang Y, Yu YZ (2015) Application of three-dimensional transitional isoparametric elements to finite element analysis of geotechnical engineering problems. Rock and Soil Mechanics 36(5):1455–1462
Feng CX, Bing L, Rui CH (2014) Finite Element Method and Its Projects, Chinese edn. Science Press, Beijing
Lin TJ, Li RF (1995) 3-D transition isoparametric element and its stress smoothing method in non-linear structural analysis. Computational Structural Mechanics and Applications 01:116–121
Liu YR, Zhou WY, Yang Q, Chen X (2005) Parallel 3D finite element analysis and its application to hydraulic engineering. Chin J Rock Mechan Eng 24(14):2434–2438
Mehl S (2006) Use of Picard and Newton iteration for solving nonlinear ground water flow equations. Groundwater 44(4):583–594
Miller CT, Abhishek C, Farthing MW (2006) A spatially and temporally adaptive solution of Richards' equation. Adv Water Resour 29(4):525–545
Phoon KK, Toh KC, Chan SH, Lee FH (2002) An efficient diagonal preconditioner for finite element solution of Biot's consolidation equations. Int J Numer Meth Eng 55(4):377–400
Sonneveld P, van Gijzen MB (2008) IDR (s): A family of simple and fast algorithms for solving large nonsymmetric systems of linear equations. SIAM J Sci Comput 31(2):1035–1062
Wang XC (2003) Finite element method. Tsinghua University Press, Beijing
Williams GA, Miller CT, Kelley CT (2000) Transformation approaches for simulating flow in variably saturated porous media. Water Resour Res 36(4):923–934
Wu MX, Gao LS (1999) Saturated unsaturated unsteady seepage numerical analysis [J]. Journal of Hydraulic Engineering 12:38–42
Zhang YL, Feng XT, Ru ZL (2004) Large-scale high performance parallel finite element system based on domain decomposition method in geomechanics. Chin J Rock Mechan Eng 23(21):3636–3641
Zheng C, Zhang JH (2007) SSOR-PCG method used in simulation of geotechnical engineering with finite element method. Chin J Rock Mechan Eng 26(S1):2820–2826
Zhu BF (2009) The finite element method theory and applications. China Water & Power Press, Beijing
Zienkiewicz OC, Taylor RL (1977) The finite element method, 3rd edn. McGraw-hill, London

Chapter 6
Seepage and Consolidation Analysis of High Earth-Rockfill Dam

The Nuozhadu high earth-rockfill dam is the highest earth-rock dam in China. Its seepage-consolidation characteristics are very representative of high earth-rock dams. A complete monitoring system was installed in the dam, and a great deal of field observation data have been obtained. In this chapter, the field observation data of pore-water pressure in the core wall of the Nuozhadu high earth-rockfill dam is first analyzed. Then, the finite-element program developed in this dissertation is applied to the analysis of unsaturated consolidation of the dam. Comparing the numerical results with the observation data, the reasonability of the results is proved, and the multi-physics coupling mechanism in the core wall of high earth-rockfill dams is explained. Moreover, the applicability of the theory and program established in this dissertation to the large-scale three-dimensional consolidation analyses of complex high earth-rockfill dams is also discussed.

6.1 Description of the Nuozhadu High Earth-Rockfill Dam Project

The Nuozhadu hydropower station is located on the mainstream of the lower Langcang River, near the Pu'er City of Yunnan Province. This project composed of the earth-core rockfill dam, an open spillway on the left bank, flood discharge tunnels, and underground water diversion and power generation structures. The total installed capacity is 5850 MW, and the designed annual average power output is 2.39×10^{10} $\text{kW} \cdot \text{h}$. The earth-rockfill dam with a maximum height of 261.5 m is the highest earth-rock dam that has been built in China. This project is a milestone in the construction of ultra-high earth-rockfill dams.

Y. Wu, *Multi-physics Coupling Analysis of Clayey Core Wall of High Earth-Rockfill Dam*, Springer Theses, https://doi.org/10.1007/978-981-13-1032-4_6

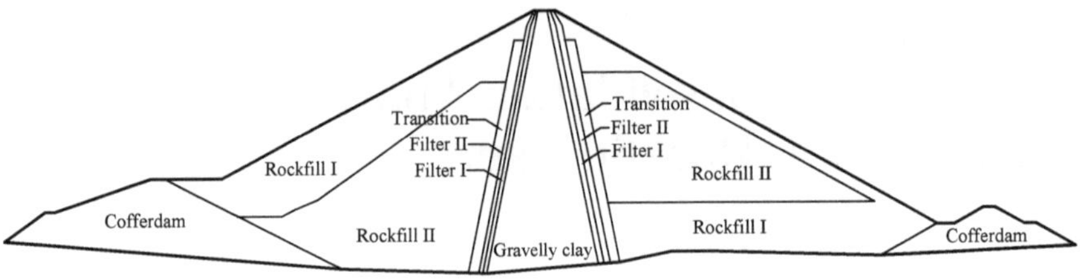

Fig. 6.1 The maximum cross-section of the Nuozhadu high earth-rockfill dam

6.1.1 Overview of the Dam Material Zoning and Construction

The maximum cross-section of the dam, with material zoning, is shown in Fig. 6.1. There are six types of dam materials in total, which are gravelly clay (i.e., the core material), rockfill I, rockfill II, transition, filter I, and filter II. The crest elevation of the Nuozhadu high earth-rockfill dam is 824.1 m, and the normal water level of the upstream reservoir is 812.0 m. The upstream slope of the dam is 1:1.9, the downstream slope of the dam is 1:1.8, and the slopes of the upstream and downstream faces of the core wall are both 1:0.2. To reduce the uncoordinated deformations between the core wall and the rockfill, a certain amount (i.e., 35%) of gravel is mixed into the clay to increase the modulus of the core material.

The construction process of the core wall of the Nuozhadu dam and the impounding process of the upstream reservoir are shown in Fig. 6.2. Dam construction was started in January 2009 and completed at the end of December 2012. During the construction period, the filling of the dam was suspended in the flood season every year. The upstream water level was stabilized at approximately 605 m before December 2011. Then, the water rose with the impounding of the reservoir. At the end of December 2012 (construction completed), the upstream water level reached 774 m. Figures 6.3 and 6.4 show the construction process of the upstream and downstream rockfill zones, respectively. In the early stage of dam construction, there are relatively significant differences in the filling progress among different material zones. And in the late construction stage, the construction progress of different material zones is nearly the same.

6.1.2 Layout of the Monitoring Instruments and Typical Monitoring Data

The Nuozhadu high earth-rockfill dam is the highest earth-rock dam in China. To ensure safety, a large number of monitoring instruments have been installed in the dam. During the construction and operating periods, a mass of timely and comprehensive filed monitoring data have been collected to analyze the behaviors of dam

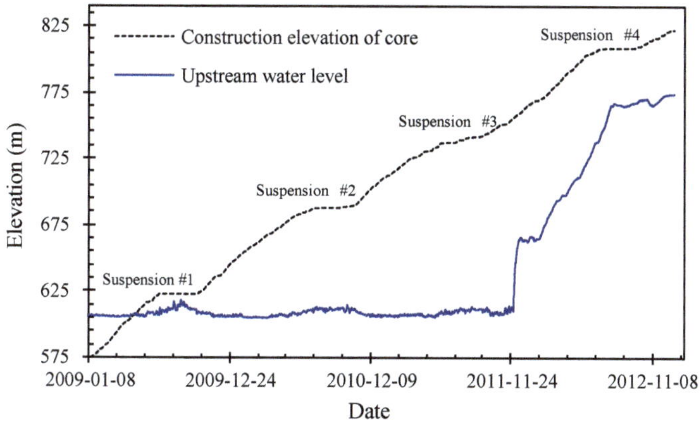

Fig. 6.2 Construction process of the core wall and water level of the reservoir

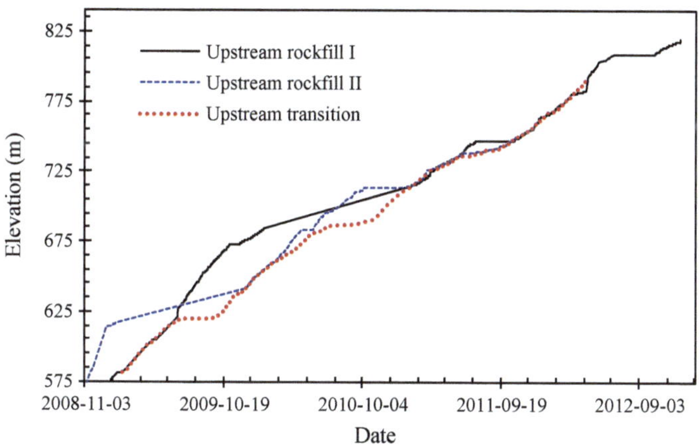

Fig. 6.3 Construction process of the upstream rockfill materials

materials and the stress, deformation and seepage characteristics of the dam body, and then provide the basis for dam safety assessment.

Monitoring instruments for pore water pressure were installed in the maximum cross-section of the dam to monitor the seepage status of the core wall during the construction period and impounding period. The arrangement of the osmotic pressure gauges is shown in Fig. 6.5. These instruments were installed at five different elevations: 626, 660, 701, 738, and 780 m. Among them, the monitoring points DB-C-P14, 24, 33, 42, and 49 are located in the upstream filter I, the monitoring points DB-C-P 18, 28, 37, 46, and 53 are located in the downstream filter I, and the other monitoring points are located in the core wall.

During construction and impounding periods, the pore-water pressure in the dam was monitored by the osmotic pressure gauges. The pore-water pressure data of the

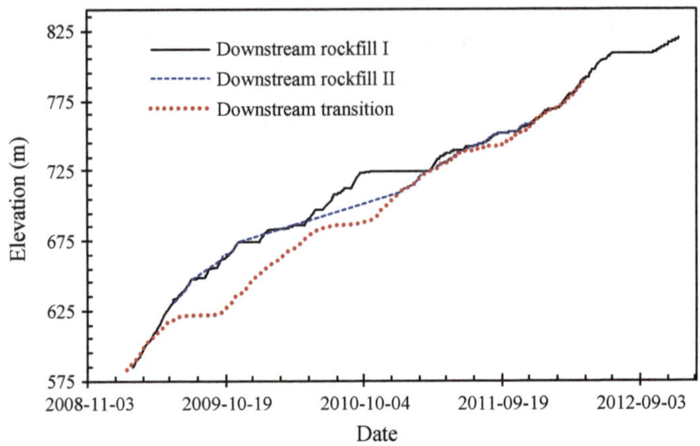

Fig. 6.4 Construction process of the downstream rockfill materials

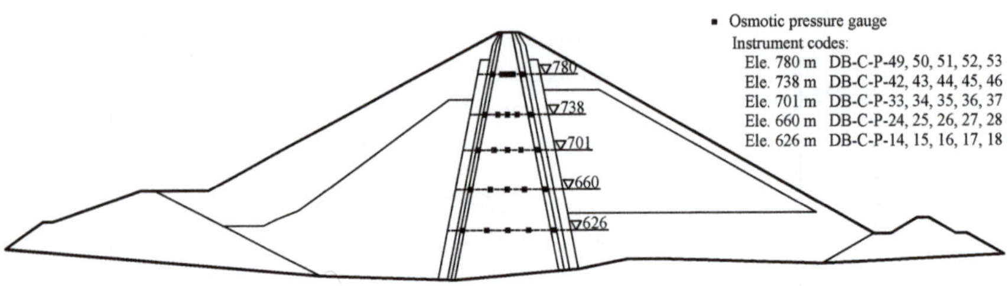

Fig. 6.5 Arrangement of the osmotic pressure gauges in the maximum cross-section

monitoring points at different elevations are shown in Figs. 6.6, 6.7, 6.8 and 6.9. In the figures, the pore-water pressure is expressed by the form of the total head (i.e., the sum of the pressure head and the potential head) to compare the observation data with the upstream water level more conveniently and to show the excess pore-water pressure more intuitively. In addition, the construction process of the core wall and the upstream water level are also plotted in the figures.

It can be seen from the observation data that the total head at all the monitoring points (except the final segment data of DB-CP-27) is higher than the upstream water level. The total head at the monitoring points of elevation 626 m even exceeds the filling level of the core wall during the whole construction period, which indicates that a high excess pore-water pressure exists in the core wall. In addition, the trends of the total head in the core wall are basically consistent with the dam construction process: the total head increases with the increase of the filling level of the core wall; when construction suspended, the total head trends to be stable; as the dam construction restarts, the total head also starts to increase gradually.

With the filling of the dam, the load on the dam increases continuously, and the soil skeleton is gradually compressed. In this process, the pore water and pore air in the core wall cannot be completely discharged in time, and excess pore-water

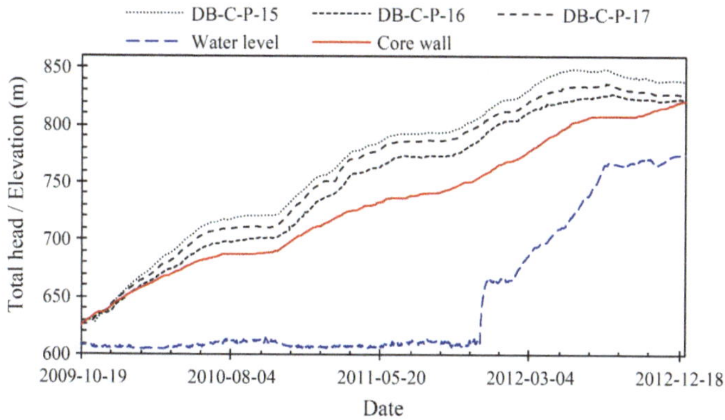

Fig. 6.6 Observation data of the pore-water pressure at elevation 626 m

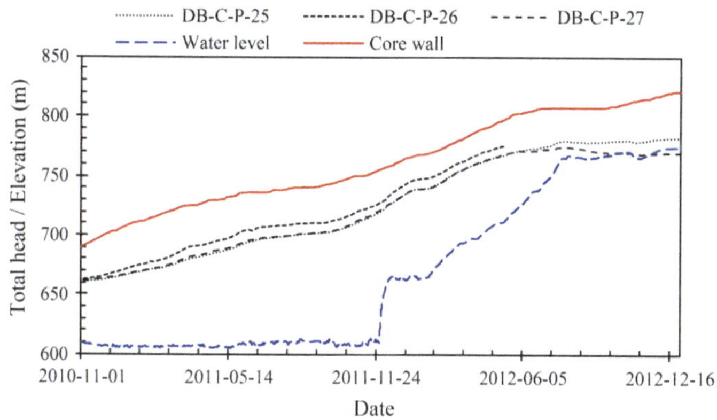

Fig. 6.7 Observation data of the pore-water pressure at elevation 660 m

pressure will be generated. Generally, the excess pore-water pressure in the core wall is mainly affected by two factors. On the one hand, the self-weight load of the dam increases continuously during construction. On the other hand, the permeability of the gravelly clay in the core wall decreases significantly due to the stress changing, the compression, etc., further reducing the dissipation rate of excess pore pressure.

Figure 6.6 shows distributions of the pore-water pressure at the monitoring points of elevation 626 m. The pore pressure at each measuring point tends to decrease at the final construction stage, indicating that the impact of construction load on the pore pressure is less than the dissipation rate of the excess pore-water pressure. With the increase of the filling level of the dam, the impact of construction load on the pore-water pressure gradually weakens. On the one hand, the width of the working face decreases as the construction of the dam, which makes the increasing rate of the load decreases. On the other hand, the distance between the location of the construction load and the monitoring points of pore-water pressure gradually increases.

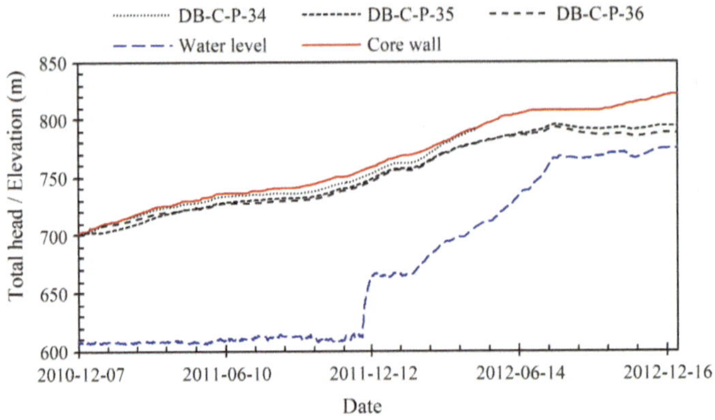

Fig. 6.8 Observation data of the pore-water pressure at elevation 701 m

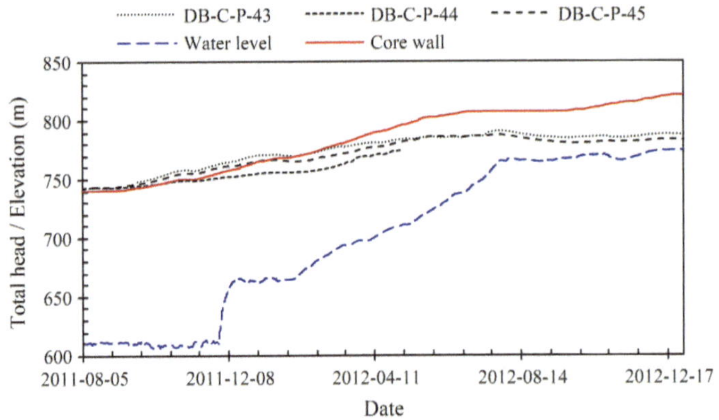

Fig. 6.9 Observation data of the pore-water pressure at elevation 738 m

In addition, as can be seen from the figures, the reservoir starts to impound water at the end of November 2011, and the rapid increase of water level has no obvious effect on the distribution of the pore-water pressure in the core wall.

The spatial distributions of the pressure head in the core wall at different elevations at the completion of the dam construction (December 18, 2012) are shown in Figs. 6.10, 6.11, 6.12 and 6.13. It can be seen that the pore-water pressure in the core wall presents an arched distribution at completion. The pore-water pressure in the middle of the core wall is higher than that near the upstream and downstream borders of the core wall. The closer to the border of the core wall, the easier the excess pore-water pressure dissipates. The leftmost and rightmost monitoring points of each elevation are located in the filter layer with relatively higher permeability. The pore pressure values at the rightmost monitoring points are 0, and the values at the leftmost monitoring points are equal to the static water pressures generated by the impoundment in the upstream reservoir. Also, the closer to the bottom of the core

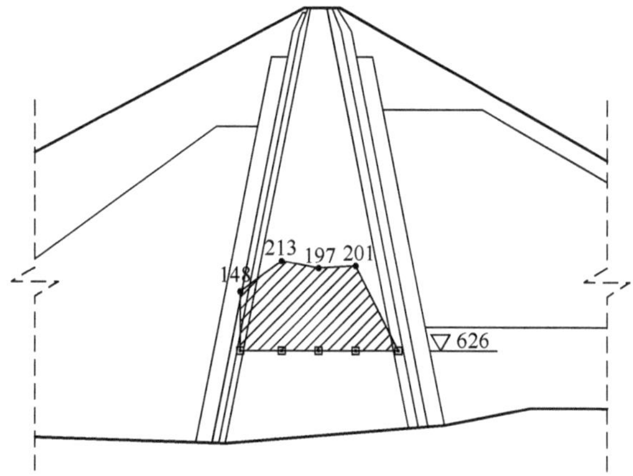

Fig. 6.10 Spatial distribution of the pressure head at elevation 626 m (2012-12-18) (*unit* m)

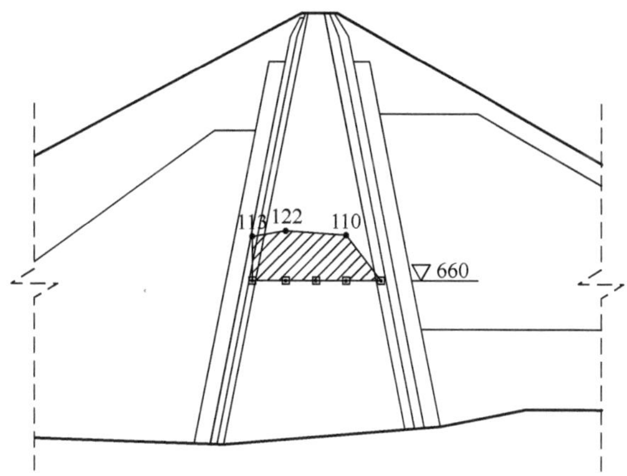

Fig. 6.11 Spatial distribution of the pressure head at elevation 660 m (2012-12-18) (*unit* m)

wall, the higher the pore-water pressure will be. On the one hand, a lower position in the core wall will have a higher static pore-water pressure. On the other hand, the stress and deformation of the core material in the lower part of the core wall are greater than that in the higher part, resulting in higher excess pore-water pressure, lower soil permeability, and also lower dissipation rate of the pore-water pressure.

6.2 Finite Element Model and Parameters

6.2.1 Finite Element Model for Numerical Simulation

A three-dimensional finite element model of the Nuozhadu dam, which contains 23,713 nodes and 23,283 elements, is constructed based on the dam site topography,

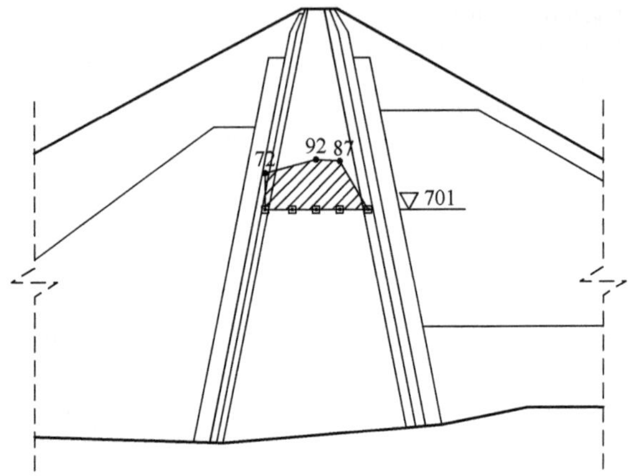

Fig. 6.12 Spatial distribution of the pressure head at elevation 701 m (2012-12-18) (*unit* m)

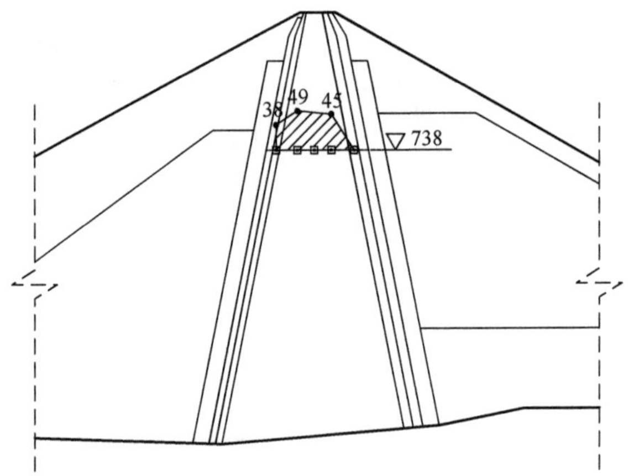

Fig. 6.13 Spatial distribution of the pressure head at elevation 738 m (2012-12-18) (*unit* m)

the dam material zoning, the construction process, and the reservoir impounding process, as shown in Fig. 6.14. Figures 6.15 and 6.16 show the maximum cross-section and the maximum longitudinal-section of the FE model. The simulated construction process of the FE model adopts the actual filling process of the Nuozhadu dam. Figure 6.17 shows the comparison between the monitored and the simulated construction process. And Fig. 6.18 shows the comparison between the monitored and the simulated impounding process. It can be seen from the figures that the 3D FE model can well simulate the construction and impounding process of the Nuozhadu high earth-rockfill dam.

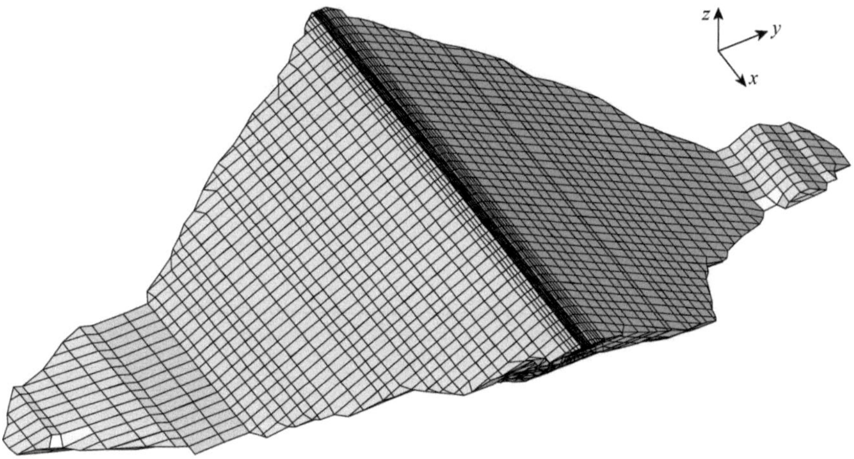

Fig. 6.14 Three-dimensional FE mesh of the Nuozhadu earth-rockfill dam

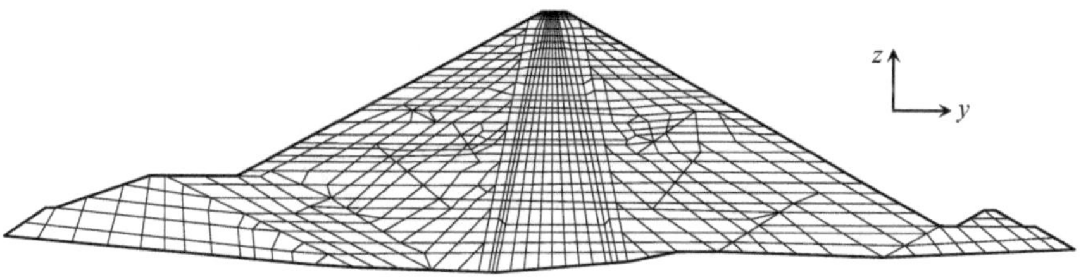

Fig. 6.15 Maximum cross-section of the 3D-FE mesh

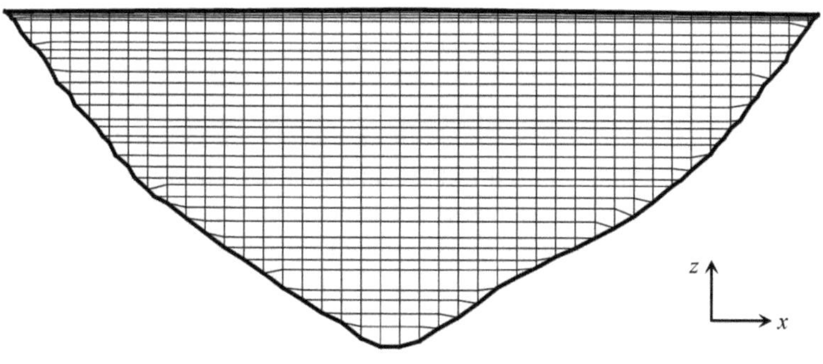

Fig. 6.16 Maximum longitudinal-section of the 3D-FE mesh

6.2.2 Constitutive Model Parameters and Unsaturated Parameters

The multi-physics coupling analysis program established in this study is applied to analyze the saturated-unsaturated consolidation of the Nuozhadu high earth-

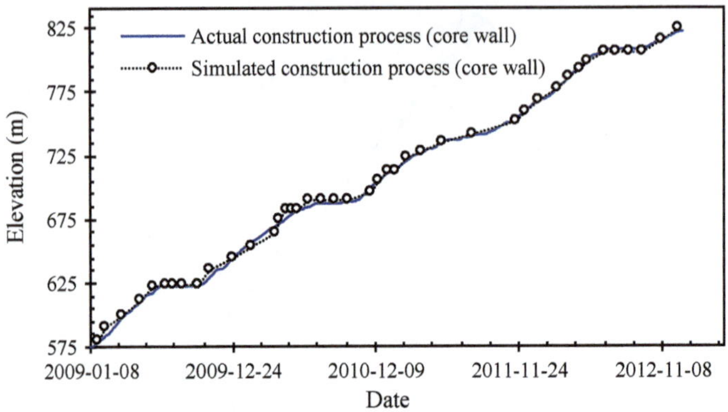

Fig. 6.17 Comparison between the monitored and the simulated construction process

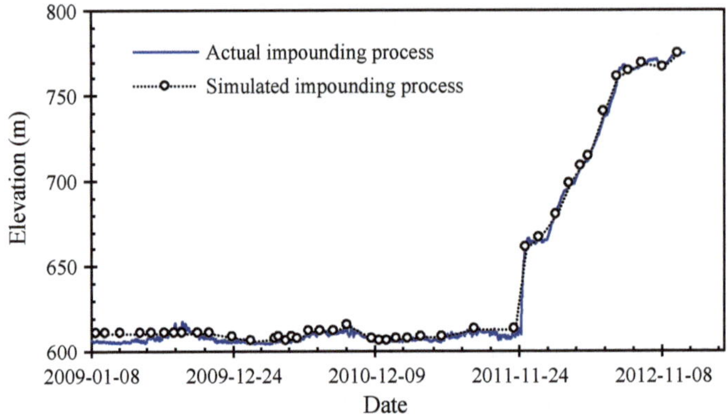

Fig. 6.18 Comparison between the monitored and the simulated impounding process

rockfill dam. The parameters used in the simulation are shown in Tables 6.1, 6.2 and 6.3. Table 6.1 shows the parameters of Duncan and Chang's EB model of the six dam materials. These parameters were obtained from the results of the previous displacement-based back-analysis based on the monitoring data of the Nuozhadu dam (Wu et al. 2014). The van Genuchten model, the parameters of which are shown in Table 6.2, is employed in the numerical simulation to describe the soil-water characteristics. Table 6.3 shows the unsaturated characteristic parameters of the gravelly clay of the core wall.

Table 6.1 EB model parameters of the dam materials

Material	K	n	R_f	K_b	m	φ_0 (°)	$\Delta\varphi$ (°)
Gravelly clay	510	0.25	0.76	340	0.15	39.34	9.8
Rockfill I	1486	0.20	0.73	665	0.10	54.20	10.1
Rockfill II	1643	0.20	0.74	717	0.10	51.31	10.2
Transition	1100	0.28	0.69	530	0.12	50.54	6.7
Filter I	1067	0.25	0.76	327	0.19	52.60	10.1
Filter II	1115	0.24	0.67	481	0.21	50.96	8.0

Table 6.2 Parameters of the soil-water characteristic curve

Material	α (m^{-1})	β	θ_r	θ_s
Gravelly clay	0.005	1.15	0.38	0.068
Rockfill I	0.05	3.15	0.38	0.045
Rockfill II	0.05	3.15	0.38	0.045
Transition	0.03	1.30	0.38	0.045
Filter I	0.03	1.30	0.38	0.045
Filter II	0.03	1.30	0.38	0.045

Table 6.3 Unsaturated parameters of the gravelly clay

Material	m_2	s_e (kPa)	C_h
Gravelly clay	0.37	40	0.02

6.2.3 Back-Analysis of the Hydraulic Conductivity Model Parameters

Hydraulic conductivity is one of the most important parameters in soil consolidation analysis, which will significantly influence the stress, the deformation, the pore pressure, and their changing. Meanwhile, variations in the physical state and mechanical state of the soil will influence the soil permeability in reverse. During construction and operation of the dam, the permeability of the core material will change significantly, and the real hydraulic conductivity cannot be obtained directly by measurement. Thus, a back-analysis method using the neural network and evolutionary algorithm is applied to obtain the hydraulic conductivity of the core material on the basis of the observation data of pore-water pressure.

Figure 6.19 shows the flowchart of the back-analysis method (Yu et al. 2007; Yuan 2003; Zhang et al. 2005), including the three main steps:

(1) Perform forward FE analysis on the training parameter sets to generate samples and then use the evolutionary algorithm and Vogl's algorithms to train and optimize the neural network.

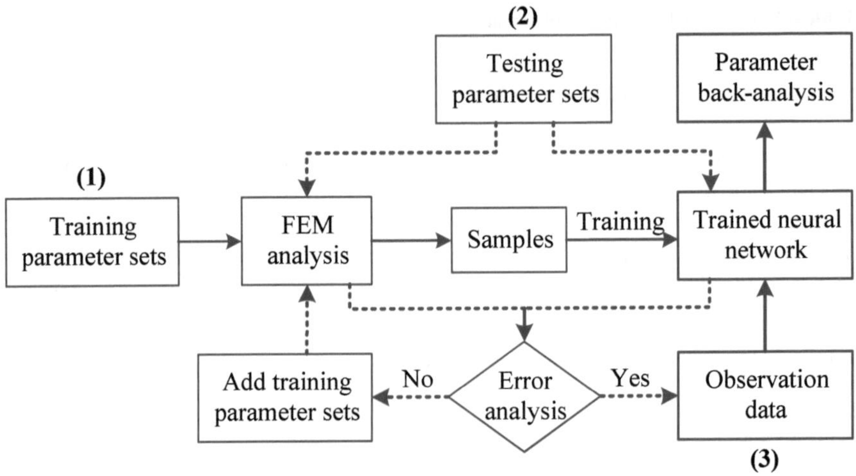

Fig. 6.19 Back-analysis method based on the neural network and evolutionary algorithm

Table 6.4 Back-analyzed parameters of the hydraulic conductivity model

Material	*a*	*b*	*c* (m/s)
Gravelly clay	43.1	−1.5	3.1×10^{-17}

(2) Access the accuracy of the trained neural network by testing parameters sets. If the error criterion is not satisfied, increase the patterns and density of the parameter set to add several more training parameter sets, and then optimize and train the neural network again until the error criterion is met.
(3) Perform parameter optimization using the observation data, the evolutionary algorithm, and the trained neural network.

The layout of the osmotic pressure gauges in the maximum cross-section is shown in Fig. 6.5. There are a total of 15 monitoring points located in the core wall at elevations of 662, 660, 701, 738, and 780 m, respectively. Reasonable observation data of the osmotic pressure gauges are used as the targets of the back-analysis, which are selected on the basis of the previous numerical analyses and data quality. In the back-analysis, field monitoring data of the points DB-C-P-16, DB-C-P-26, DB-C-P-34, DB-C-P-35, DB-C-P-36, DB-C-P-44, and DB-C-P-45 are selected. Considering the data time-histories and distributions, the observation data of DB-C-P-26, DB-C-P-34, and DB-C-P-44 on April 24, 2012, the observation data of DB-C-P-36 on June 22, 2012, and the observation data of DB-C-P-16, DB-C-P-35, and DB-C-P-45 on December 10, 2012, are selected as the targets of the back-analysis.

Considering the characteristics of the core material of gravelly clay, the parameters of the permeability coefficient model are then determined based on the back-analysis results and the numerical results of some trial calculations, as shown in Table 6.4.

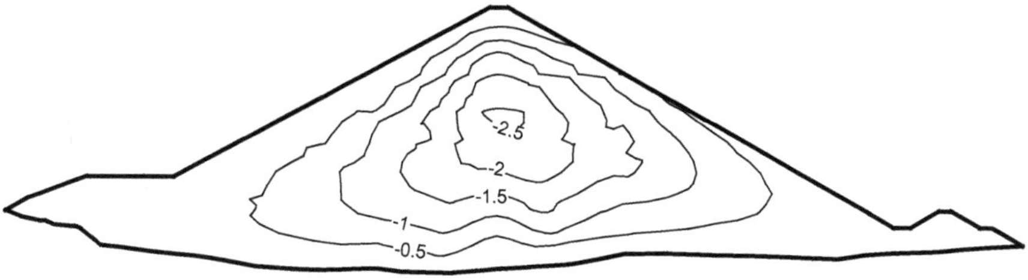

Fig. 6.20 Settlement distribution in the maximum cross-section (m)

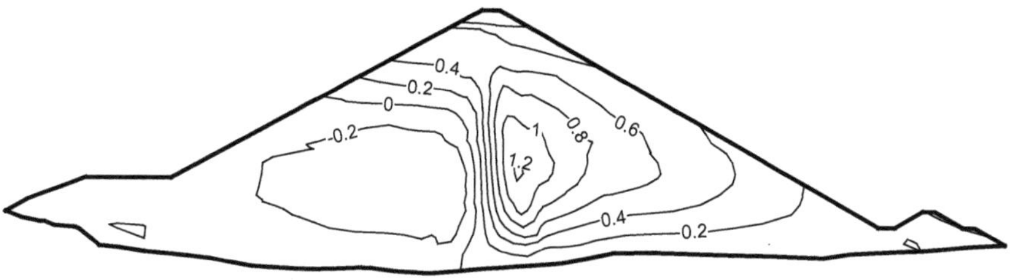

Fig. 6.21 Distribution of the horizontal displacement (*y*-direction) in the maximum cross-section (m)

6.3 Consolidation Analysis of the Nuozhadu High Earth-Rockfill Dam

With the back-analyzed parameters, the multi-physics coupling consolidation analysis of the Nuozhadu high earth-rockfill dam is performed using the FE program developed in Chap. 5. In the analyses, the wetting deformation and creep deformation are ignored, and thus the calculated deformations of the dam will not be compared with the observed data. The calculation results are as follows.

Figures 6.20 and 6.21 show the distributions of the settlement and the horizontal displacement in the maximum cross-section at the completion of dam construction, respectively. The maximum settlement is about 2.5 m, located in the center of the cross-section. And the maximum horizontal displacement is about 1.2 m. Due to the upstream water pressure of the reservoir, the horizontal displacement in most region of the dam points to the downstream direction (i.e., *y*-direction). Figures 6.22 and 6.23 show the distributions of the settlement and the horizontal displacement (along the *x*-direction) in the maximum longitudinal-section. The horizontal displacement in this section is symmetrical and points to the middle of the dam. The maximum horizontal displacement is 0.55 m.

Figures 6.24 and 6.25 show the stress distributions in the dam. The maximum values of the major principal stress and the minor principal stress are about 4 and 1 MPa, respectively. Due to the impounding of the reservoir, the effective stress in the upstream rockfill is obviously less than that in the downstream rockfill.

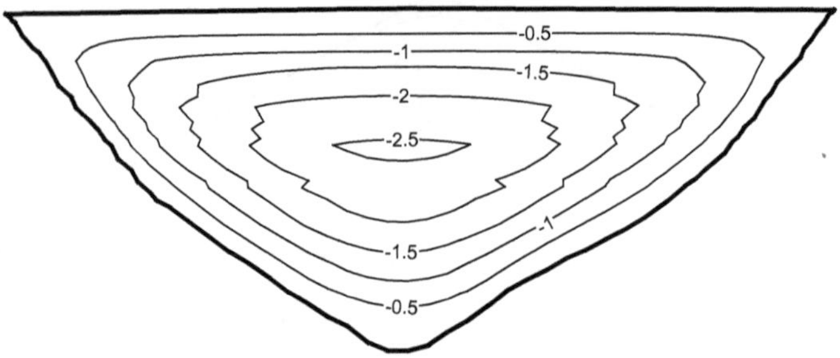

Fig. 6.22 Settlement distribution in the maximum longitudinal-section (m)

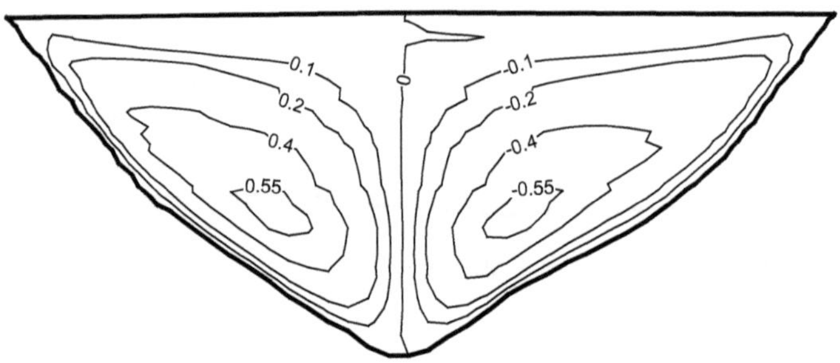

Fig. 6.23 Distribution of the horizontal displacement (x-direction) in the maximum longitudinal-section (m)

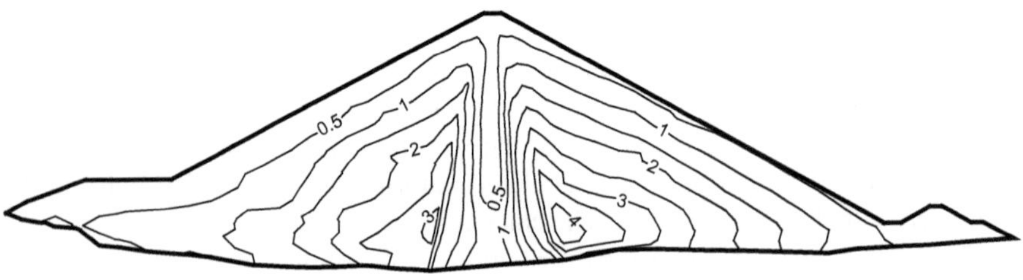

Fig. 6.24 Distribution of the major principal stress (MPa)

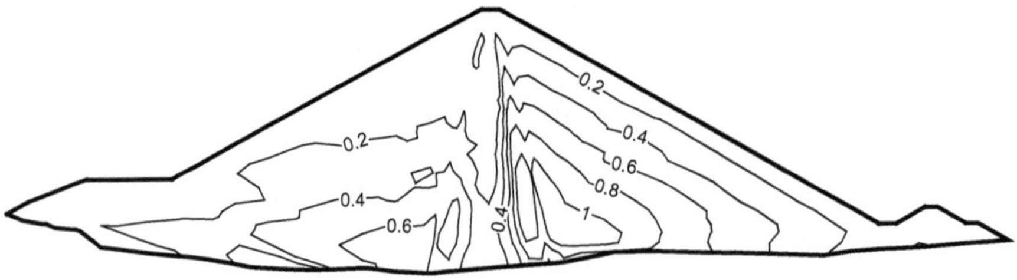

Fig. 6.25 Distribution of the minor principal stress (MPa)

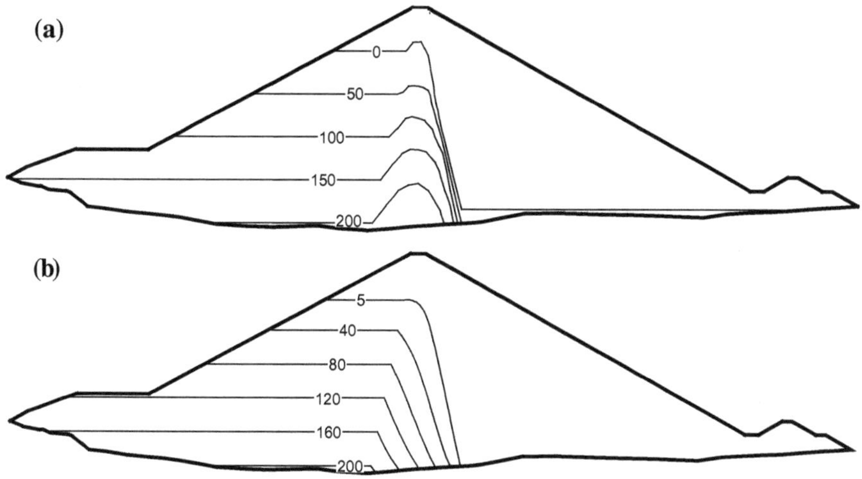

Fig. 6.26 Distribution of pore water pressure (m): **a** results calculated by the multi-physics coupling analysis program; **b** results calculated by the traditional saturated consolidation analysis program

The results of the pore-water pressure calculated by the multi-physics coupling FE program established in this study and by the traditional program for saturated consolidation analysis are shown in Fig. 6.26a, b, respectively. The two distributions of the pore-water pressure in the figures are quite different. The calculated values of the pore-water pressure in the core wall in Fig. 6.26a are higher than the values in Fig. 6.26b. Moreover, the pore-water pressure distribution of the latter is apparently different from the actual monitored distribution. The pore-water pressure distribution of the former has maximum values in the middle of the core wall and presents a decreasing trend in both sides. Also, the pore-water pressure on the left side is higher than that on the right side. It is true that this kind of distribution is due to the combined effect of impounding and dam construction.

Figures 6.27, 6.28, 6.29 and 6.30 show the comparison between the simulated total head by the multi-physics coupling FE program and the observed values at monitoring points of different elevations in the core wall until the end of 2012. Overall, the distributions of the calculated pore-water pressure coincide quite well with the observation data. For some monitoring points, although there are relatively significant differences between the observed and the calculated values, the changing trends of them are nearly the same.

Figure 6.31 shows the comparison of the spatial distributions of the pressure head in the core wall between the calculated results and the observation data at the completion of the dam. In the figures, the solid lines represent the observed values of the pressure head, and the dashed lines represent the calculated results of the pressure head using the unsaturated FE program. It can be seen that the distributions of the calculated pore pressure head coincide quite well with the observation data.

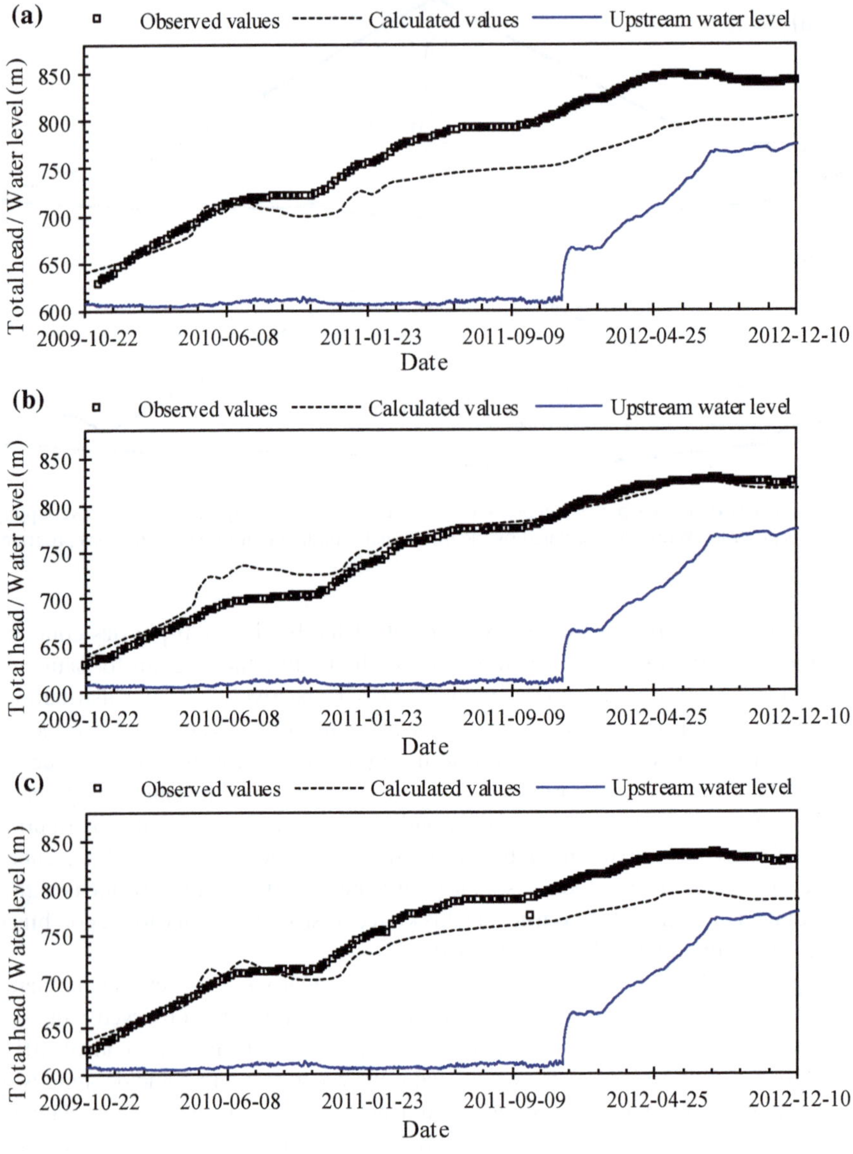

Fig. 6.27 Comparison between calculated results and observation data of water head at monitoring points of elevation 626 m: **a** DB-C-P-15; **b** DB-C-P-16; **c** DB-C-P-17

The permeability of clayey soils is significantly influenced by the physical state and mechanical state. The permeability of the core wall will change greatly during construction and impounding of the dam. For quantitative analyses, Fig. 6.32 shows the distribution of the estimated saturated hydraulic conductivity at the completion

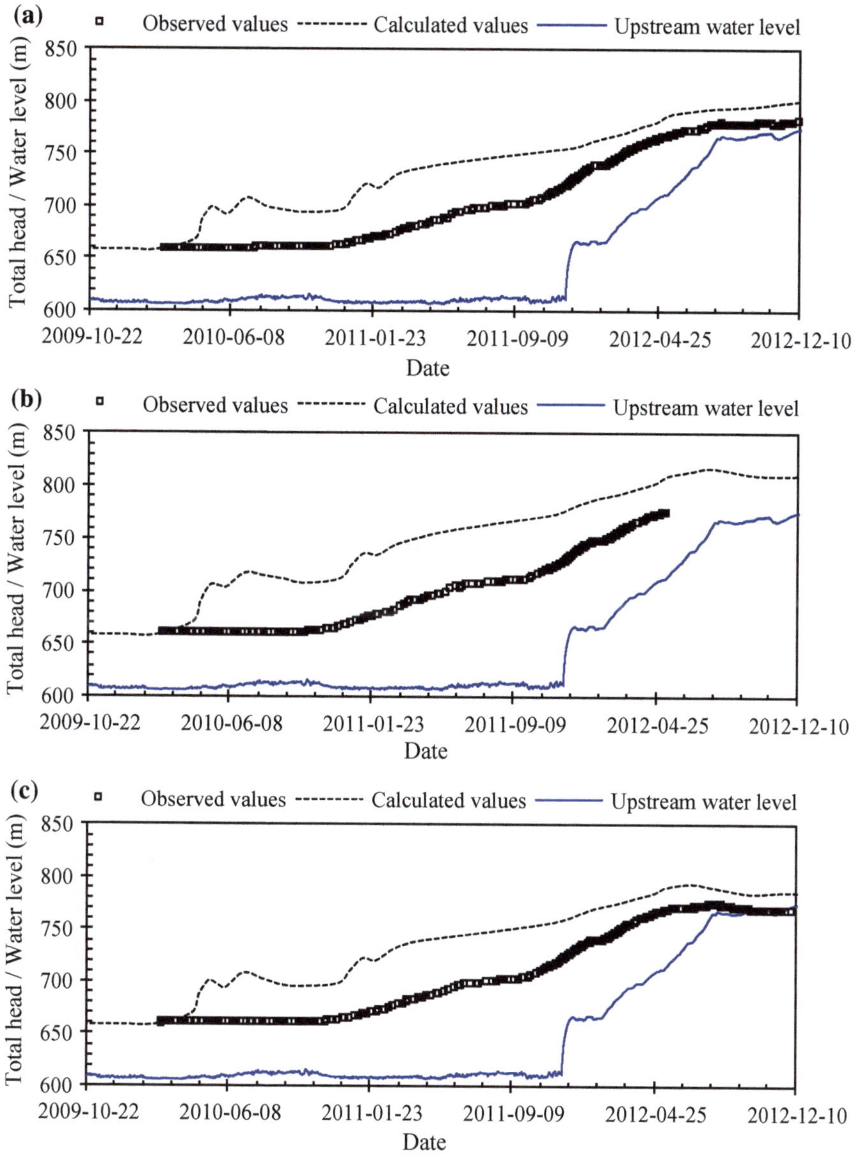

Fig. 6.28 Comparison between calculated results and observation data of water head at monitoring points of elevation 660 m: **a** DB-C-P-25; **b** DB-C-P-26; **c** DB-C-P-27

of construction (the end of December 2012). The values of the saturated hydraulic conductivity in the core wall are calculated by the multi-physics coupling FE program on the basis of the hydraulic conductivity model. It can be seen that under the influence of stress and deformation, the saturated permeability of the core wall

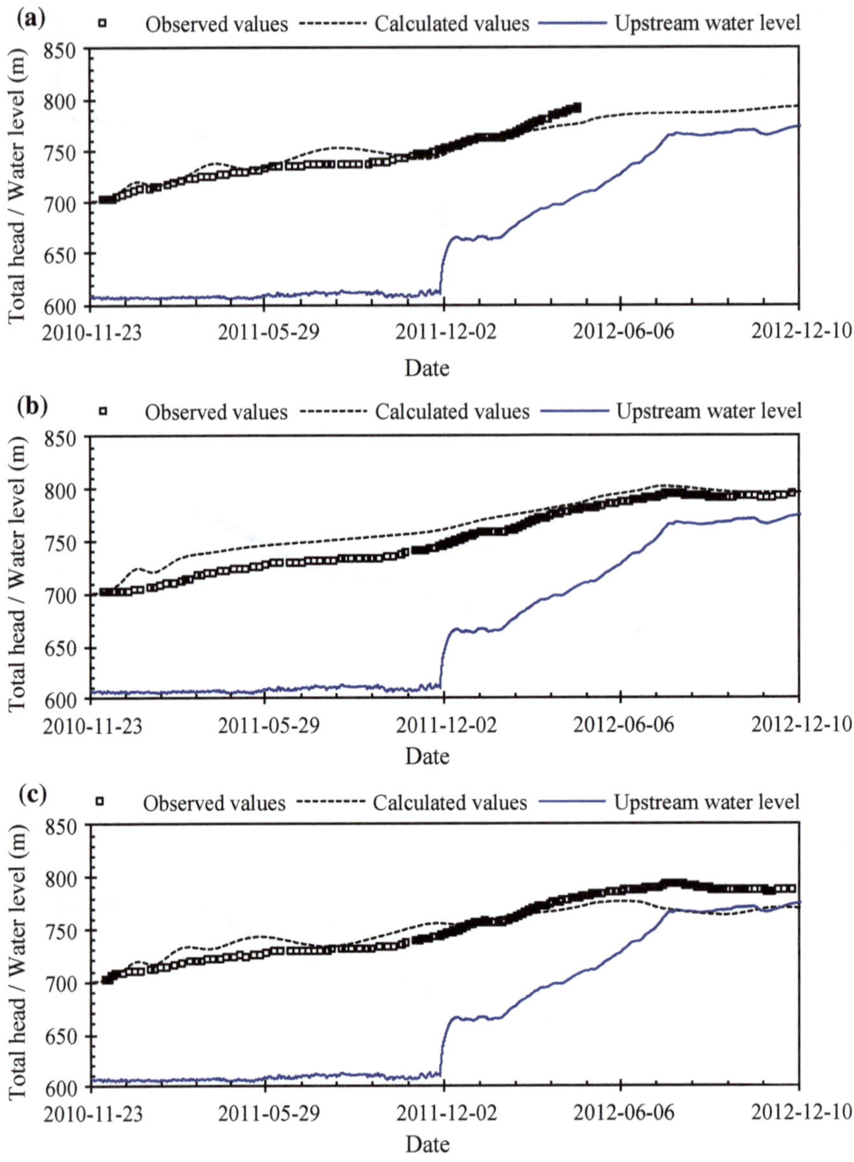

Fig. 6.29 Comparison between calculated results and observation data of water head at monitoring points of elevation 701 m: **a** DB-C-P-34; **b** DB-C-P-35; **c** DB-C-P-36

changes significantly. The hydraulic conductivity at different parts of the core wall is quite different. The hydraulic conductivity is relatively high in the top part of the core wall, with a value of about 2×10^{-8} m/s. At the bottom of the core wall, the

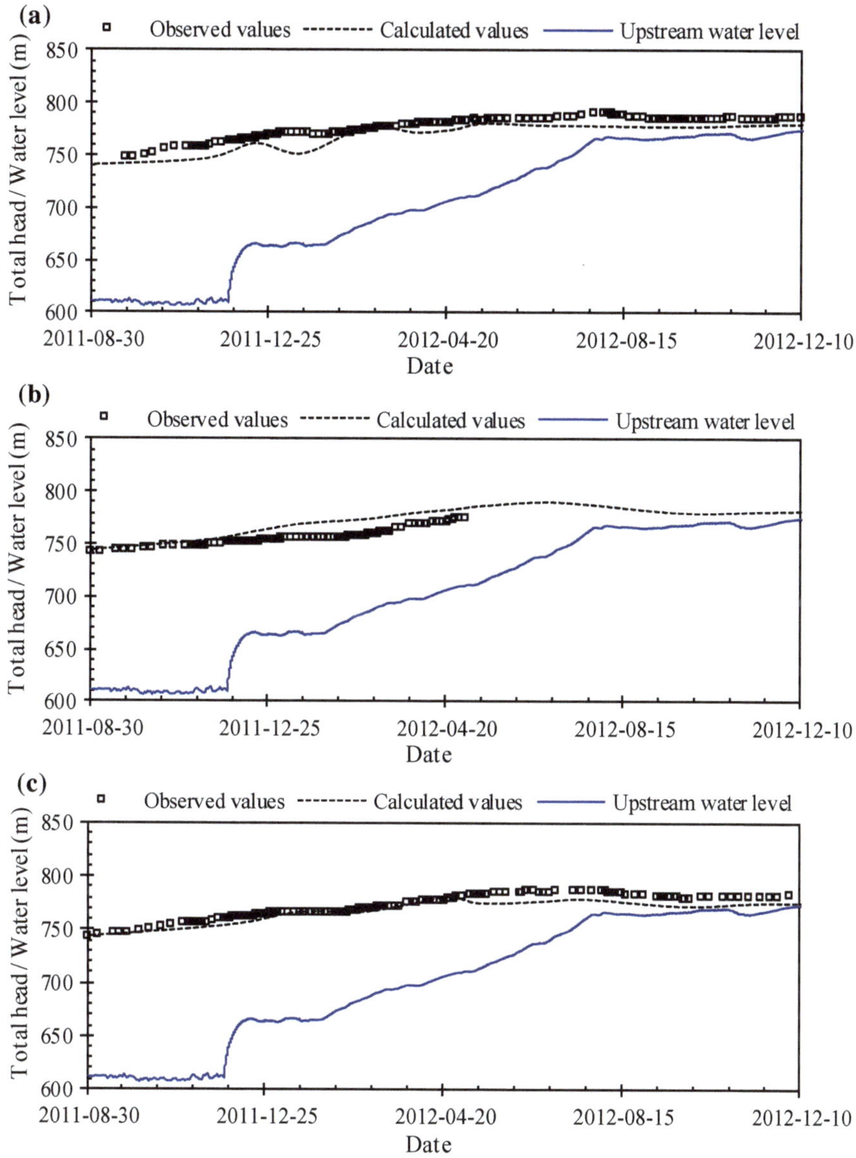

Fig. 6.30 Comparison between calculated results and observation data of water head at monitoring points of elevation 738 m: **a** DB-C-P-43; **b** DB-C-P-44; **c** DB-C-P-45

hydraulic conductivity is much smaller (about 5×10^{-10} m/s) due to the significant influence induced by the stress and deformation of the core material.

From the above, the simulation results of the Nuozhadu dam validate the coupling analysis method and the program founded in this study, which can be used to explain

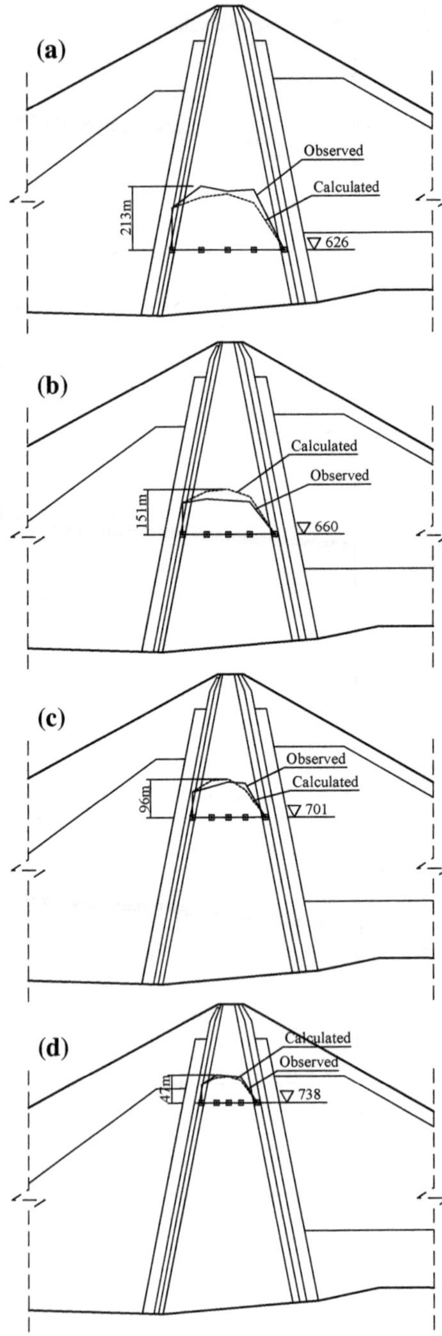

Fig. 6.31 Comparison of pore-water pressure between calculated results and observation data of different elevations: **a** elevation 626 m; **b** elevation 660 m; **c** elevation 701 m; **d** elevation 738 m

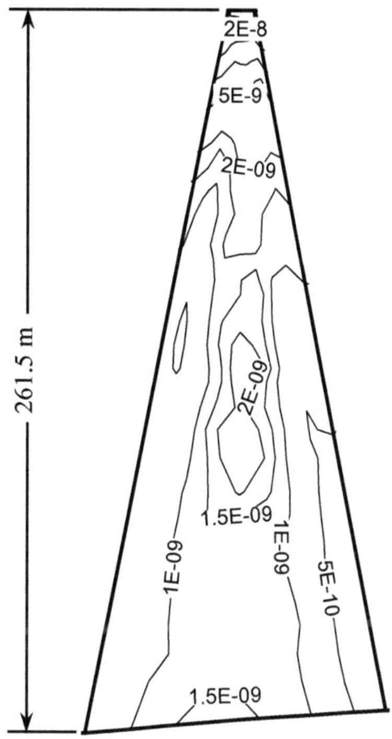

Fig. 6.32 Distribution of the estimated saturated hydraulic conductivity, *k*, in the core wall at the end of December 2012 (*unit* m/s)

the multi-physics coupling mechanism in the core wall of high earth-rockfill dams. The FE program is suitable for large-scale three-dimensional consolidation analysis of high earth-rockfill dams.

6.4 Summary

The coupling analysis method and the FE program established in this study are applied to analyze the unsaturated consolidation of the Nuozhadu high earth-rockfill dam. The Nuozhadu dam is the highest earth-rockfill dam in China. Its seepage-consolidation characteristics are very representative of high earth-rock dams. A complete monitoring system was installed in the dam, and a great deal of field observation data have been obtained. Based on the observation data and the numerical simulation result, the following conclusions can be drawn:

(1) The field observation data show that the pore-water pressure in the core wall remains at a high level and shows an increasing trend during the construction of the dam. The pore-water pressure is closely related to the dam construction process. On the one hand, the soil skeleton of the core material is compressed under the filling load, which makes the excess pore-water pressure generates. On

the other hand, the permeability of the core material reduces with the changing of the stress and deformation, which makes the dissipation rate of the pore water pressure decreases.

(2) The multi-physics coupling consolidation analysis of the Nuozhadu high earth-rockfill dam is performed, from which reasonable results are obtained. The spatial and temporal distribution of the pore-water pressure calculated by the program coincides well with the observed values, which implies the reasonableness of the coupling consolidation analysis method and the validity of the numerical simulation program. Moreover, the distribution of the hydraulic conductivity in the core wall is quantitatively analyzed. The permeability of the core material changes significantly during construction and impounding. Therefore, the multi-physics coupling effects should not be ignored in the consolidation analysis for high earth-rockfill dams.

(3) The coupling analysis method proposed in this study can be used to explain the multi-physics coupling mechanism in the core wall of high earth-rockfill dams. The efficiency and accuracy of the analysis program are evaluated by instance analysis, indicating that the FE program is suitable for large-scale three-dimensional consolidation analyses of high earth-rockfill dams.

References

Wu YK, Yuan HN, Zhang BY, Zhang ZL, Yu YZ (2014) Displacement-based back-analysis of the model parameters of the Nuozhadu high earth-rockfill dam. Sci World J

Yu YZ, Zhang BY, Yuan HN (2007) An intelligent displacement back-analysis method for earth-rockfill dams. Comput Geotech 34(6):423–434

Yuan HN (2003) Displacement back-analysis of embankment dam based on neural network and evolutionary algorithm. M.S. thesis, Tsinghua University, Beijing

Zhang BY, Yuan HN, Li QM (2005) Displacement back analysis of embankment dam based on neural network and evolutionary algorithm. Rock Soil Mech 26(4):547–552

Chapter 7
Conclusions and Future Work

In recent years, with the growing social demand and the development of construction technology, earth-rockfill dams built in China are being built higher and higher. In high earth-rockfill dams, the seepage and consolidation behaviors of the clayey core wall are significantly different from those of low dams, showing apparent multi-physics coupling effects. These phenomena are difficult to reasonably explain and accurately simulate with previous theories and methods. Therefore, in this dissertation, on the basis of the previous studies, the mechanical properties of unsaturated compacted clayey soil, the multi-physics coupling consolidation theory, the numerical programming, and their practical engineering applications to high earth-rockfill dams are investigated. The main work and conclusions of this dissertation are as follows.

7.1 Conclusions

(1) Saturated and unsaturated triaxial CD tests are conducted to study the mechanical properties of the clayey core material of the Nuozhadu high earth-rockfill dam. The mechanical properties of unsaturated soil are significantly different from those of saturated soil. With the same test conditions, the strength and modulus of the unsaturated soil are higher than those of the saturated soil, especially for the results of low confining pressures. When the soil is unsaturated, the matric suction in the soil plays an important role in the effective stress, which will significantly affect the strength and deformation properties of the soil. With the increase of the confining pressure, the shrinkage of the soil specimen increases, and correspondingly, the degree of saturation of the unsaturated soil increases. Hence, the effect of matric suction on the effective stress gradually weakens. The mechanical properties of the core material will change obviously with saturation during construction and impounding so that the role of matric suction should be fully considered in the stress-deformation analysis of unsaturated soils.

Y. Wu, *Multi-physics Coupling Analysis of Clayey Core Wall of High Earth-Rockfill Dam*, Springer Theses, https://doi.org/10.1007/978-981-13-1032-4_7

(2) On the basis of the simplified consolidation theory for unsaturated soils, a consolidation analysis method for unsaturated soil considering the effects of multi-physics coupling is established. In this method, some core issues, such as the calculation of the effective stresses, the modeling of the soil permeability, and the constitutive models, are emphatically explained; and the interactions between the physical-state field, the stress-deformation field, and the seepage field are fully considered. This method gives a more rational and comprehensive seepage-consolidation analysis mode for saturated and unsaturated soils. Moreover, the unsaturated effective stress parameter of the clayey core material in the Nuozhadu high earth-rockfill dam, which can be applied to the unsaturated consolidation analysis of the dam, is obtained based on the data of the triaxial CD tests.

(3) The finite element equations of the multi-physics coupling analysis method for seepage and consolidation are derived by using the Galerkin method. And the initial conditions and boundary conditions in the seepage-consolidation analysis of earth-rock dams are analyzed and explained. Based on the established coupling analysis method and the finite element equations, a program for unsaturated seepage and consolidation analysis is developed. Moreover, efficient solvers for large-scale linear equations, a nonlinear iterative algorithm, and an under-relaxation method are introduced into the FE program to improve the accuracy, efficiency and computational stability. Then the program is adopted to conduct unsaturated seepage consolidation analyses of two ideal examples, and reasonable numerical results are obtained. Comparing the results of the FE program with those of a traditional saturated consolidation program, it can be seen that the matric suction in the unsaturated soil has significant effects on its stress and deformation characteristics in unsaturated consolidation analyses. Compared with the traditional saturated analysis program, the multi-physics coupling analysis program developed in this study has significant advantages. For example, the newly developed program can perform unsaturated seepage and consolidation analyses; it considers the changes of soil permeability with the stress, deformation, and physical state; the effects of the matric suction on the stress and deformation are taken into account; and by using a nonlinear iterative solution in the program, the accurate seepage-field results (e.g., distributions of head and phreatic line) can be obtained.

(4) The multi-physics coupling analysis program is applied to analyze the seepage and consolidation behaviors of the Nuozhadu high earth-rockfill dam. The numerical results of stress and deformation are reasonable. And the calculated spatial and temporal distributions of the pore-water pressure in the core wall agree well with the observed values, which validates the rationality of the proposed coupling analysis method. Moreover, the distribution of the permeability coefficient in the core wall is quantitatively analyzed. The hydraulic conductivity of the core material changes significantly during construction and impounding. Hence, the multi-physics coupling effects should not be neglected in the consolidation analyses for high earth-rockfill dams. On the whole, considering the computational accuracy and validity of FE program, the program is suitable for large-scale three-dimensional analyses of complex high earth-rockfill dams.

7.2 Future Work

The seepage and consolidation problems of unsaturated soils widely exist in geotechnical engineering. The multi-physics coupling analysis method for unsaturated seepage and consolidation problems proposed in this study explains the principle of multi-physics interaction mechanism in high earth-rockfill dams. This method will have broad application prospects. The work in this dissertation promotes the practical application of the unsaturated seepage-consolidation theory and the multi-physics coupling analysis method. However, due to the complexity of the unsaturated seepage and consolidation analysis in high earth-rock dams and the limitations of the work in this study, there are still some issues to be further studied:

(1) In the numerical solutions of the coupling analysis, more efficient and stable nonlinear iterative algorithm could be introduced into the FE program of this study. The numerical solutions are the necessary way from theory to practical engineering applications. In this dissertation, the Newton-Raphson nonlinear iterative method was adopted in the FE program. Although the combination of this method and the efficient iterative solutions for linear systems of equations have met the demand of large-scale complex engineering, there is still room for improvement in the computational efficiency of the nonlinear iterative solutions.
(2) In the unsaturated consolidation theory, the values of the unsaturated parameters in consolidation analysis could be studied more deeply. In numerical calculations, the rationality of the parameter values directly determines the correctness of the results. Due to the fact that the unsaturated consolidation theory has rarely been used in practical engineerings at present and few pieces of information are available for reference, the values of some unsaturated parameters (e.g., the air drainage ratio) are still determined by experience and need to be further studied.
(3) More accurate constitutive model parameters could be obtained by back-analysis method. In the consolidation analysis of the Nuozhadu high earth-rockfill dam, due to the massive workload of this study, the constitutive model parameters were taken from the results of the previous displacement-based back-analysis based on the monitoring data of the Nuozhadu dam. In order to simulate the stress-deformation characteristics more accurately, the constitutive model parameters and the permeability model parameters of the clayey core material can be back-analyzed together based on the field observation data of displacement and pore-water pressure.
(4) The application of unsaturated seepage-consolidation theories in the consolidation analysis of high earth-rockfill dams is scarce. More research needs to be done in combination with typical large-scale engineerings to verify the rationality of the theory and the results further.

MIX
Papier aus verantwortungsvollen Quellen
Paper from responsible sources
FSC® C105338

If you have any concerns about our products,
you can contact us on
ProductSafety@springernature.com

In case Publisher is established outside the EU,
the EU authorized representative is:
Springer Nature Customer Service Center GmbH
Europaplatz 3, 69115 Heidelberg, Germany

Printed by Libri Plureos GmbH
in Hamburg, Germany